OMNICONOMICS

Omniconomics shows how we can make human society intrinsically sustainable, harmonically embedded in nature, with the help of a completely new approach in which traditional economics is transformed.

Citing the fallacies of existing economic approaches as directly responsible for many of the environmental and social threats faced by society today, Niko Roorda presents a new, interdisciplinary science: omniconomics. This framework has its foundations in an innovative, joined-up approach, in which all aspects of the natural and social sciences are inextricably linked. It will better allow for new solutions to tackling urgent issues, including climate change, deforestation, environmental and economic inequality, dehumanization, and crumbling social cohesion.

Inspiring new thinking, this book aims to:

1. Inspire a public, scientific and political debate about the role of present-day economics, its status as no more than a protoscience, and the consequences for (inter)national policies and developments;
2. Stimulate independent, out-of-the-box thinking of students, graduates, politicians, leaders of society, and any member of the public who worries about the major present-day problems and the future;
3. Stimulate interdisciplinary research within universities towards the creation of an omniconomic science that will replace the discipline of economics and will become a genuine, mature science;
4. Challenge the readers to go beyond short- or medium-to-long-term solutions such as the Sustainable Development Goals and circular economy: *Omniconomics* starts where *Doughnut Economics* ends

Encouraging the reader to rethink accepted economic approaches to deal with the present-day threats of global catastrophes, this book will be of great interest to students and scholars of sustainable development, sustainability economics and policies, as well as any practitioners working in related fields.

Niko Roorda, PhD, has been working for more than thirty years on integrating sustainable development into education, businesses, and society: as a lecturer, curriculum developer, program manager, project manager, auditor, author, consultant and teacher trainer. A theoretical physicist and philosopher of science by origin, he had his PhD on the topic of sustainability, higher education and organizational change. His academic textbook, *Fundamentals of Sustainable Development*, has been published in four editions in three different languages. In 2001, Roorda received the Dutch National Award for Innovation and Sustainable Development for his achievements and in 2018, he was selected as the Sustainable Lecturer of the Year in the Netherlands.

Author website: https://niko.roorda.nu

Omniconomics downloads:
https://niko.roorda.nu/books/omniconomics/#Downloadlinks

OMNICONOMICS

The Re-Creation of Economics for a Sustainable Future

Niko Roorda

Routledge
Taylor & Francis Group
LONDON AND NEW YORK

earthscan
from Routledge

Cover image: Niko Roorda

First published 2023
by Routledge
4 Park Square, Milton Park, Abingdon, Oxon OX14 4RN

and by Routledge
605 Third Avenue, New York, NY 10158

Routledge is an imprint of the Taylor & Francis Group, an informa business

Initially published in Dutch as *Omniconomie. De weg naar een intrinsiek duurzame wereld* by Gompel&Svacina.

British Library Cataloguing-in-Publication Data
A catalogue record for this book is available from the British Library

ISBN: 978-1-032-30687-2 (hbk)
ISBN: 978-1-032-30689-6 (pbk)
ISBN: 978-1-003-30626-9 (ebk)

DOI: 10.4324/9781003306269

Typeset in Bembo
by SPi Technologies India Pvt Ltd (Straive)

Access the Support Material: www.routledge.com/9781032306872
and https://niko.roorda.nu/books/omniconomics/#Downloadlinks

CONTENTS

Foreword *ix*

1 The Problem and the Plan **1**

PART 1

The Problem: Economics is not a Science **5**

2 Toward the roots of unsustainability **7**

2.1 Dig deeper 7

2.2 Narratives 12

2.3 Laws and regulations 16

3 Four Spheres of thoughts and actions **19**

3.1 Words, narratives, policies, effects 19

3.2 Interventions in the Four Spheres 26

4 Intrinsic sustainability **33**

4.1 Glued-on versus intrinsic sustainability 33

4.2 The Broad Holloway 35

5 Impetus words: Concepts that need to be thrown away **42**

5.1 The birth of science 42

5.2 17th century, physics and astronomy: impetus 44

5.3 18th century and beyond: chemistry, biology, medicine, neuropsychology 46

5.4 Economics: value, profit, externality, ownership etc. 49

6 Protoscience **58**
6.1 The seven characteristics of protoscience 58
6.2 Schools 60
6.3 Pattern recognition and illusions 64

7 Economics myths **74**
7.1 Automatisms 74
7.2 More myths 77
7.3 Copied methods 79
7.4 Oversimplification 83
7.5 Lack of replicable experiments 85
7.6 Economics is a myth 87
7.7 Conclusion: economics is a protoscience 90

8 The unsustainable success of economics **91**
8.1 Scientific success 91
8.2 Growth 93
8.3 Profit 101

9 A bitter success: *Homo Solitarius*, the Lonesome Human **109**
9.1 Alienation from nature 110
9.2 Alienation from each other 116
9.3 Alienation from society 122
9.4 Conclusion 127

10 Collapses that may bring our final demise **129**
10.1 Growth patterns 129
10.2 Theoretics of catastrophe 132
10.3 First signs 135
10.4 The final demise 141

11 The inevitable conclusion: proto-economics **145**
11.1 Seven characteristics 145
11.2 Economics, magic, religion, protoscience 152

PART 2

The Plan: Omniconomics, the new Science **155**

12 Buying time with the SDGs and more **157**
12.1 The SDGs & the Earth Charter 157
12.2 System inertia 161

13 The concept of understanding **165**
13.1 Connecting the dots 165
13.2 Connecting upside down 176
13.3 Connecting upside down in economics 182

14 Characteristics of science **190**
14.1 Magic, Religion, Philosophy, Science 190
14.2 Philosophy of science 194
14.3 Values-based science 199
14.4 Aggregation levels 201

15 The eightfold path to science **207**
15.1 New physical equipment 208
15.2 New mental equipment 208
15.3 New patterns 209
15.4 New words 211
15.5 New narrative 212
15.6 New successes 213
15.7 New questions 214
15.8 New fire 216

16 Characteristics of omniconomic science **221**
16.1 Scientific criteria: coherent, reflective, methodical, empirical 222
16.2 Systemic criteria: complex, complete, multilevel, unifying 223
16.3 Sustainability criteria: intrinsic, dynamic, strategic, peaceful 224
16.4 Engagement criteria: values-based, inclusive, purposeful, accountable 225

17 The complexity of the omniconomic world **228**
17.1 New equipment: complexity 228
17.2 New words: homeostasis, attractors 232
17.3 New equipment: simulations 234

18 Tegular unification **243**
18.1 New equipment: network theory 243
18.2 Level transitions 244
18.3 The rooftile structure 248

19 On the road to success: evolution towards sustainability **251**
19.1 New equipment: self-learning systems 252
19.2 New narrative: evolution towards intrinsic sustainability 257
19.3 Survival of the Sustainablest 261

20 Participatory democracy: society directs science **267**
20.1 Support 268
20.2 Inclusivity 272
20.3 Personal message to the reader 279

Acknowledgments, Credits, Literature, Appendices *281*
Notes *283*
Index *292*

FOREWORD

by Prof. Jan Peter Balkenende, former Prime Minister of the Netherlands; Professor of Governance, Institutions and Internationalisation at Erasmus University Rotterdam; External Advisor to EY

It is a surprising new vision that Niko Roorda offers in this book. His answer to the major problems we face in our time is: redesign the world economy from scratch. Even better: redesign the underlying economic science first, because that is not yet a science at all – as the author argues in a sometimes humorous, sometimes vicious manner. His book reads like a thriller.

The book shows that the contemporary problems that threaten us, from climate change to collapsing stock exchanges and rapidly rising inequality in wealth and power, have their deepest roots in the global economic structure. He uses numerous examples to show that this stems from a number of fundamental misconceptions – myths, the author calls them – about economics, in a way that I find convincing.

According to Roorda, the current work on sustainability, including SDGs and the circular economy, and including his own work of more than thirty years, is therefore not sufficient, because that is glued-on sustainability, while intrinsic sustainability is required, which is in the genes of society.

It would be easy to leave it at that, but Roorda's ambition goes further: he also comes up with solutions. He proposes to extend economic science – protoscience, according to the author – to omniconomics, which includes nature alongside the economic world, as well as human values that cannot be expressed in money: beauty, friendship, respect.

To discover how this can be done, the author examines how other sciences, including his 'own' physics (Roorda is both a graduate physicist and philosopher, and a PhD social scientist), have come about: with the help of new instruments of perception, new ways of thinking, new 'narratives'. Currently, we have a wide range of new methods and ways of thinking at our disposal, such as complexity, computer simulations, and artificial intelligence.

While it is the task of scientists to create the new science in an interdisciplinary way, it is up to society to define the goals of that science. Inclusiveness and participation are key words here.

The path to intrinsic sustainability outlined in the book has the potential to contribute significantly to finding solutions to the problems of our time. It is good that this book has been written: it inspires and gives hope to all who are concerned about our shared future on this beautiful planet.

Jan Peter Balkenende

1

THE PROBLEM AND THE PLAN

We just don't quite understand what we're doing yet. We often think we have put together a well-thought-out economic system but, in reality, the global economy just sort of spontaneously emerged. Without purpose, without design, without understanding. The economy is a tinkerer's job; we don't even understand how it works. That's the problem. That's why so much is going wrong in the world.

Things can be different. Better. I describe a plan for doing so in this book.

We do not yet understand the fundamental connections in the economy. Because all sorts of things are going wrong, we tinker and turn all the knobs, but it doesn't help. Because we don't know what to do to create a stable, safe and sustainable world. We just can't get the hang of the economy. Economics is still far from being a science, and so every economic policy amounts to muddling through. To groping in the dark.

Economic theorists would love to build their field with as much success as the natural sciences, which have achieved remarkable results over the past few centuries. Physics, chemistry and biology have constructed grand theoretical edifices that have greatly enhanced our understanding of physical reality. This has produced tremendous successes. Modern technology is able to land robots on other planets. To connect all people online, to cure or even prevent numerous diseases, to fight pandemics, and to produce enough food for at least ten billion people.

At the same time, the economy fails to get even the most basic things done. As a result of our economic activity, the natural environment is being systematically destroyed. Animal and plant species are dying out en masse. In oceans, dead zones are growing. The climate is changing, land and polar ice is melting, seas are rising, even in this century world cities are going to flood. Refugee flows of hundreds of millions of people seem inevitable. While a minority of humanity enjoys unprecedented prosperity, and while there is more than enough food to fill all mouths,

DOI: 10.4324/9781003306269-1

hundreds of millions of people are in the direst circumstances of poverty, malnutrition and lack of security, health care and education. Rich people are rapidly becoming richer; exploding inequality is undermining solidarity in society. And still, from time to time the global economy collapses in chaotic fashion as bubbles burst and stock markets collapse.

While technology has entered the 21st century, the economy is still in the Stone Age.

Our incapacity, in twenty-five schools

Solutions to make the economy more successful are offered from all sides. More than twenty-five schools of economics are fighting with each other over the most basic topics. If physics and astronomy were still at that same primitive stage, astronomers would be arguing today about how to infer the future from the position of the planets.

At the same time, we are working, with some success, to make the world more sustainable. But the sustainability we are currently achieving is not changing the world in a fundamental way, because we don't yet know how that could be done. Because we don't understand our own economic and political processes and therefore can't redesign them. As a result, current sustainable development leads to ***glued-on sustainability***: that is, sustainability for which we – or rather, future generations – will have to make constant efforts to maintain it. That kind of sustainability is unstable, it falls over as soon as you stop your efforts, because it is not fundamentally integrated into the global system. And that is not good enough: what we need is a form of sustainability that is stable in itself. So that it stays upright even if we don't pay attention to it. I call that: ***intrinsic sustainability***. It is the goal of my plan and of this book, and it must become the goal of sustainable development and, in fact, of all of society.

And so, the first, necessary step of the plan is: to recognize that economics as a discipline is not yet remotely a science. To admit our inability.

Economics is nothing more than a ***protoscience***, which mixes some of the characteristics of incipient science with properties that rather belong to religion, to philosophy, to magical thinking and to flat superstition. In the first part of this book, I will tell what that is, a protoscience, and prove beyond any doubt that economics really is still such a protoscience. After which I show how incredibly damaging and dangerous that is, since numerous governments, administrators and boards do not realize that the 'scientific facts' they use as the basis for their decisions and their strategic policies are in reality fairy tales. Myths. Narratives told using words that are flawed because they steer our minds in the wrong direction.

First understand, then build

The logical next step of my plan is: to point out a path along which a real, mature science can be developed. I will make it clear that such a science cannot be just about economics, just as – say – car technology cannot be just about the chassis but must be about the whole car and even about roads, traffic laws and people's behavior. Similarly, the new mature science to be developed should be not only about economics but also about sociology, psychology, politics, human rights, environment, nature, sun and wind and much more. This new science, which does not yet exist, must have a name: I call it ***omniconomics***, a word that begins with 'omni-': everything.

I hope you don't expect me to design this omniconomics all by myself. Of course, this is impossible: this science can only be developed if scientists from all sorts of disciplines cooperate intensively in an interdisciplinary way. Better still: if society at large is also involved, science and society can together decide what kind of world, what kind of society, the new omniconomic models should produce.

Although I cannot develop the omniconomic science on my own, I can show which road we can take to make it happen. This is the crux of my plan, for which I will carefully study the developments that in past centuries led to the fields of science that are so successful today. I will show that important elements in the creation of new sciences were: the availability of new instruments, of new mathematical or logical techniques and models of thought, and even of new words. Examples from the past were: the telescope, the clock, and probability theory.

Right now, in the 21st century, a tremendous amount of such innovative elements is available. These include: computers, the Internet, big data, complexity, network theory, self-learning systems, artificial intelligence, virtual reality, artificial life, and many more. Together, these are going to enable us to create the omniconomic science so desperately needed.

So that we understand what we are doing. So that we can purposefully build a global human system in which sustainability is an intrinsic feature of human society, obviously harmoniously embedded in the natural society of our planet.

Download: appendices, literature and resources

You have a conclusive narrative in your hands here: the book is complete. Yet there is so much more I would like to tell you. Juicy details. Additional examples. A scientific or philosophical underpinning. Or – admittedly – hobbyhorses. You don't need them for the narrative in this book, but if you'd like to delve deeper into the subject, they are extremely interesting.

For that purpose, I have written a large collection of ***appendices***, which together actually form a complete additional book that you can read without cost. From the main book (i.e., this one), I regularly refer to these appendices.

To find them, go to https://niko.roorda.nu/books/omniconomics: there you will find the links where you can download the appendices.

Perhaps I'll write more such appendices in the future – not referred to from the book, of course, because that will already have been printed. But I'll mention them on the website.

Even better: who knows, maybe you, as a reader, will be so fascinated by a subject in the book that you will want to do further (scientific or practical) research yourself and write an appendix about it yourself, which then – if I agree – will be put on the website, including your name of course. I would love that. If you are planning something like this, always contact me in advance.

Via that same web page, you can also look up the ***Bibliography***. The sources to which I refer in the text by means of numbers such as this,[286] are not in the printed book itself, which means that the book can remain thinner and may therefore cost less.

Furthermore, in a number of places I offer you ***tools*** that enliven the narrative. Such as movie clips, spreadsheets, and computer programs that you can use to experiment all by yourself. The links to these tools can also be found on the web page I mentioned.

I wish you much reading pleasure and fascination, and perhaps wildly enthusiastic or indignant passion.

Kind regards,
Niko Roorda
Sprang-Capelle, Netherlands
https://niko.roorda.nu/contact
March 2022

Recommended further reading (for details, consult the online Bibliography)

Book: Roorda, Niko (2021): *Fundamentals of Sustainable Development.* Routledge, 3rd edition.

Part 1. The *Problem*:

Economics is not a Science

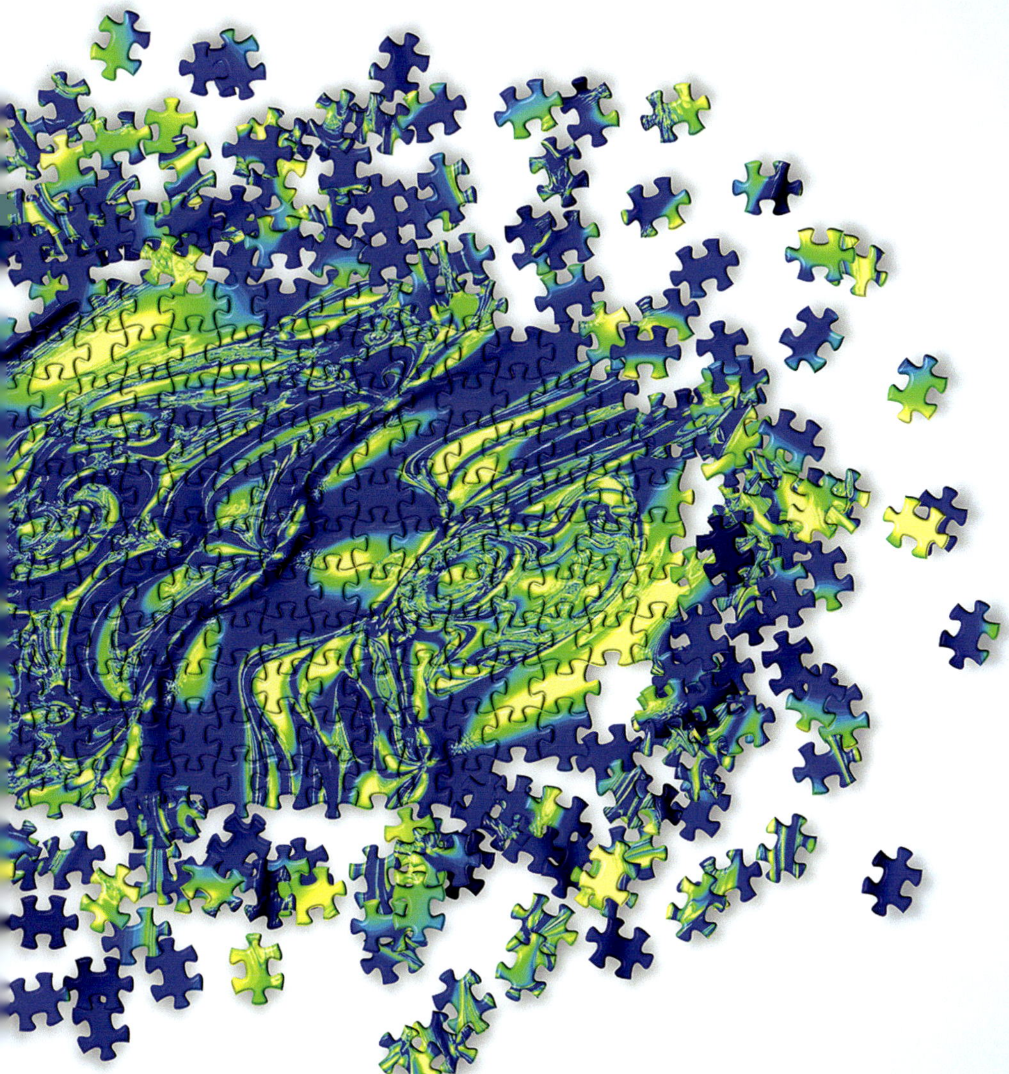

2

TOWARD THE ROOTS OF UNSUSTAINABILITY

2.1 Dig deeper

It's great. No, really. The efforts people spend on sustainability. And on corporate social responsibility, CSR. More and more people are working on it now, and they work hard. Not only individual people, but also governments, companies and civil society organizations.

But it is not enough.

I have been working fanatically on sustainability for more than thirty years, with all my passion, energy and ideals. In those years I have seen the efforts grow, as more people take the sustainability themes seriously – with the climate problem at the forefront.

About ten years ago, sustainability started to become a hype, and that has only become stronger ever since. A hype, not only in the negative sense – a buzzword – but also in the positive: sustainability is now taken very seriously. The newspapers write about it every day; in politics, it has become a hot issue. Every year, billions of euros and dollars are invested sustainably.

And yet it is not enough.

It needs to go deeper. More thorough. More fundamental. Because despite all these efforts we do not tackle the root causes of the current unsustainability, with all its major disasters and threats.

In this book I tell why. And I don't stop there, because otherwise the book would not be very meaningful. I will also describe ways in which we can make our world sustainable from its fundaments. *Intrinsically sustainable.* I will explain that term, it is the core of the book.

I start with a true story in which an absurdity turned into a global disaster. It is a story from a financial-economic angle; I start with an economic topic because economics is central to this book. After the description of its aftermath, you will

DOI: 10.4324/9781003306269-3

see how this story has to do with unsustainability with regard to 'profit', which, together with 'people' and 'planet', is generally regarded as one of the three main aspects of sustainable development.

A beginning of an end

It started with an oil spill disaster. And it ended with the largest financial and economic catastrophe in world history.

It was March 24, 1989. Along the frozen shores of Alaska entered a supertanker loaded with crude oil. It was the *Exxon Valdez*, owned by oil company Exxon.

The huge ship was en route from the port city of Valdez, the location of the southern terminus of the Trans Alaska Pipeline, heading for California. But it would never arrive, because in an icy bay, Prince William Sound, the tanker struck a reef on the seabed, the Bligh Reef. The ship ripped open and a large amount of oil ended up in the sea: the estimates range from 42 to 144 million liters (11 to 38 million gallons).

In the days and months after, a shoreline length of 2100 kilometers (1300 miles) was contaminated, and a sea surface of 28,000 square kilometers (11,000 square miles) was heavily polluted.

Environmental organizations and local residents – mostly fishermen – were, to put it mildly, not pleased. And millions of marine animals and birds will not have been happy either, even though you cannot ask them: the bay was an important ecological area. The scale of the disaster was huge, especially because, at the very low temperatures in Alaska, oil takes years to degrade.

The state of Alaska was obviously not happy too. Exxon was sued in what would become a long-term lawsuit, and Alaska demanded a compensation of US$5 billion. Even for the mighty Exxon, this was a large amount. Therefore, the oil company was looking for ways to hedge against this demand. They approached J.P. Morgan & Co., one of the largest banks in the United States, and managed to secure a loan of US$4.8 billion.

Even though J.P. Morgan had provided the loan, the bank was not really assured that the repayment would ever take place. What would happen if the claims after the disaster caused Exxon serious financial difficulties? Perhaps the loan could not, or could only partially, be reimbursed, making the oil company 'in default'. The House of Morgan, as the bank is called, searched for a way to reduce risk.

One of the bank employees found a solution. 'Can you perhaps *insure* against the risk of no or poor repayment – a *credit default* – of the loan?', she wondered. The answer was 'yes': the European Bank for Reconstruction and Development was willing to take out such insurance. This European investment bank took a share of the risk incurred – the risk swapped ownership – in return for which J.P. Morgan paid a premium.

The 'Credit Default Swap' was born. In 1994, fourteen years before the major financial explosion.

The rise

Such a Credit Default Swap (CDS) proved an excellent plan. There were more banks – and other investors – who thought it was a great idea to insure their outstanding loans to unreliable borrowers.

J.P. Morgan saw a gap in the market. The bank brought the CDS on the business market as a new *derivative*, i.e., a new investment vehicle, itself with no immediate value but holding a *derived* value, based on the value of other matters. In this case, the value of funds by outstanding loans.

Trade in CDSs grew rapidly, especially after 1997, when J.P. Morgan made it possible for smaller investors, for whom billion-dollar investments were far too large, to buy small CDS packages. Soon, all kinds of CDSs were traded, including ones linked to bonds issued by countries such as Greece.

By 1998, the CDS portfolio of J.P. Morgan had grown to US$50 billion. But that was only a small part of the trade. Other banks now also traded in CDSs, and the overall CDS market in that year was estimated at $300 billion.

Estimated. Because there was a snag. Unlike some other derivatives, CDSs are not traded in publicly accessible and controlled exchanges but directly between the parties involved. That was not all, because CDSs were not only sold by lenders, but over time they were mutually traded by various parties. Thus, the CDS market was cluttered and almost invisible and therefore uncontrollable. Transparency was completely absent

This became really important when some banks invented something new. Why – they wondered – should you be the provider of a loan or bond to base a CDS on it? They started selling credit default swaps at a rapid pace to anyone who wanted them, even to institutions that had never provided risky loans. These remarkable new products were called *Naked Credit Default Swaps*. The naked CDSs were no insurances at all: they were just gambling systems. They were also *negative* gambling systems, because the owner of a naked CDS made a big profit when a company or a country that had received the loan was in financial difficulties. Look at this example, in which a fictional land and fictitious banks occur – except J.P. Morgan.

Numerical example

The Scandinavian country Friesmark issues a government bond worth a total of €400 million. A bank named Broad Bank buys these bonds. For safety, Broad Bank buys a CDS for half of their value at J.P. Morgan. Because Friesmark is seen as a financially safe country (AAA rating), the premium is low: Broad Bank pays an annual premium of €3 million, in exchange for which Morgan guarantees the insured amount of €200 million.

This is all perfectly fine: the CDS is decently based on a package of bonds Broad Bank actually owns. No nakedness here.

But now, other parties enter the stage. The hedge fund Zeros decides to buy a CDS of the same size, also based on the bonds of Friesmark. Only: Zeros holds no bonds of Friesmark! Zeros also pays €3 million per year for this naked CDS, and is now, like Broad Bank, entitled to €200 million if Friesmark cannot repay its debt.

From that moment, Zeros has every interest in ensuring that the economy is bad in Friesmark. And it does not stop at Zeros, because a dozen other parties close just such a deal with J.P. Morgan as Zeros, so a few years later up to €2.2 billion of naked CDSs have been traded: much more than the value of the bonds ever issued by Friesmark.

The owners of these naked CDSs now 'lie in wait' for a possible economic crisis in Friesmark. It cannot even be ruled out that they do more than passively lying and waiting, as they might perhaps actively lend a hand to make it difficult for Friesmark, for instance by encouraging companies to settle out of Friesmark or even leave the country. A kind of 'match fixing'; but by all means legal.

What if Friesmark indeed gets the worst of it and cannot repay the debt? Then J.P. Morgan suddenly has to cough up €2.2 billion on naked CDS value, plus another €200 million for Broad Bank. If, in the same period, the House of Morgan has a few more such setbacks, that might be enough to topple the mega bank. After which Broad Bank, not receiving its insured money, may fall over too. After which the customers of Broad Bank ...

Ever heard of the domino effect?

Not everyone is happy with this bizarre derivative. The fact that there are institutions, large and powerful institutions, that make huge profits because other parties get into crisis, is seen as dangerous. The comparison can be made with a fire insurance that you take out, linked not to your own house but to that owned by someone else. If you do that, it is not illogical that you will then hope that a fire breaks out in the other building. What would you do if, on a dark night, you saw a small beginning fire in that house?

A naked CDS is a *perverse* economic tool: it encourages antisocial behavior.

The fall

It was swinging business, those CDSs. Between 2001 and 2007, the outstanding capital doubled every year, by 2007 reaching a value of US$62 trillion. (Note the language difference: in the UK, this may be indicated as $62 billion. We talk about US$62 million million dollars; $\$62 \times 10^{12}$). See Figure 2.1.

In the same year 2007, the United States Gross Domestic Product (GDP) was $14.5 billion, less than a quarter of this...

Eighty percent of the CDS value sat in naked CDSs and so was uncovered!

The CDSs were traded among their buyers. This caused their commercial values to fluctuate. When it went bad with the underlying company or country

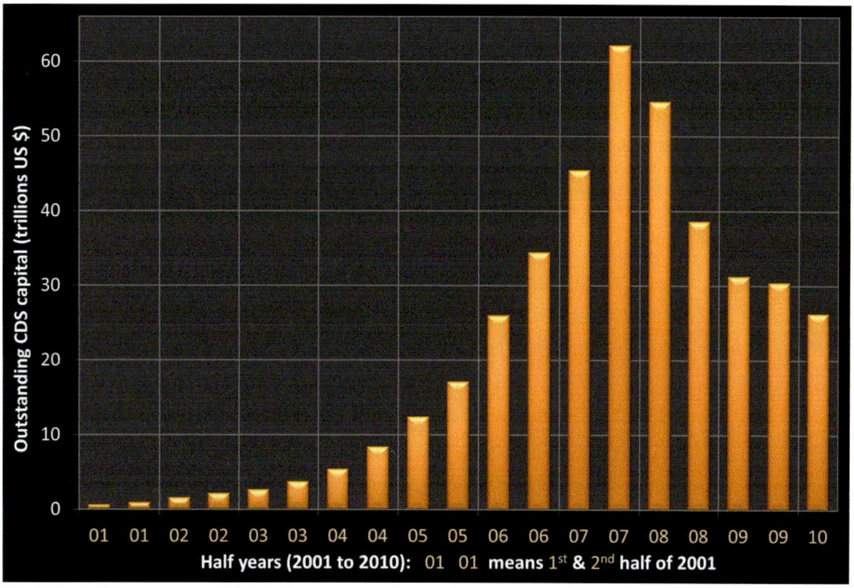

FIGURE 2.1 Outstanding CDS capital (nominal value), 2001–2010.[1]

(Friesmark, in the example), the commercial value of the linked CDSs increased, and vice versa. It was striking that the CDSs acted very nervous. Small changes in the valuation of the underlying company or country immediately resulted in large fluctuations in the market value of the CDSs. Thus, the trade in credit default swaps contributed to turbulent financial markets.

Then came the banking crisis in 2008, put in motion by the housing crisis in the United States. Here is what happened. Large packages of unsafe ('subprime') mortgages on US homes were insured by the banks through CDSs. When one after another homeowner was unable to pay the mortgage, many CDSs had to be paid. For several banks, this was too much, and they fell over, dragging other banks down in their fall, which in their turn caused others… and so on.

The CDS system was not the only reason for the housing crisis, but it was certainly an important factor in causing a devastating banking crisis that went around the world, calling a long economic malaise felt by everyone.

The question

The key question surrounding this whole development is this.

How did it happen that a method that was designed to create financial security, turned to be a major cause of financial disaster?

The CDS method was the tip of a complex construction of investment products, which together were called 'securitization': creation of security. But the CDS method caused no security at all, but rather the opposite. However, the designers of the system, experts of the mega bank J.P. Morgan, were no fools. If only that were so! Then we could have blamed them. On the contrary: the developers represented the very best that economics had to offer, and their product had been widely embraced.

What's wrong with economics, when the experts can be so wrong, raising global disaster? Don't they know what they are doing?

2.2 Narratives

Unfortunately, indeed these experts do not know what they are doing. This is why.

Naked CDSs are curious things. In fact, they are *narratives*. Narratives told by *bankers*, who believe their own narratives so strongly that others do, too.

The narrative starts with money: in itself a curious thing, because you can't eat money and you can't build houses with it. Yes, you can *buy* food and building materials with it, but that's not the same. And you can *pay* mason and carpenters who build a house for you. They do that because they believe in the curious 'money' thing, just like you. Sometimes the faith weakens, and then you need more and more money for your food or your home. As in Zimbabwe in the early 21st century, see Figure 2.2.

If there is sufficient faith in money, then you can then make people believe in even stranger things. In *banks*, which not only have a lot of money but can even create it from scratch while you stand by and watch: that is, when you *borrow*, for example for a *mortgage*: another two words that only work if people believe in them.*

FIGURE 2.2 Zimbabwe, 2008: one hundred trillion dollars. Worth: 30 dollars, or 25 euro.

* It may sound strange, but it is really true: the vast majority of money in circulation is not created by governments or central banks but by private, commercial banks (McLeay et al 2014; Bollen, 2017; WRR, 2019). This is not about *circulating currency*, i.e., printed banknotes or minted coins, but about *transferable money*, i.e., numbers on bank accounts, as the creation of currency is and remains the prerogative of governments or central banks authorized by governments. For a long time, books and journals (e.g., Tobin, 1963) have shown that private banks make the lion's share of the money, and yet this is still known to very few people (Sustainable Finance Lab, 2014).

Such a mortgage, made from money that has just been fantasized for you, can then be used to purchase a house – which really exists, where you can warm it comfortably and beget and raise your children. The purchase means that you then are the *owner* of that house: another phenomenon that only works because everyone believes in it. And if you want to keep that house safely, then you also buy *insurance*. This is a truly strange concept. A *fire insurance policy* does not guarantee that no fire breaks out, while a *life insurance policy* does not ensure that you stay alive: on the contrary, because it is certain that you will not.

Well, and for everyone who has been able to believe this fairy tale in its entirety, the narrative is still not over. Because then one day a tanker breaks open and the oil flows out. And again, a magic word appears, because then the owner of that tanker pays *compensatory damages*. That word is similar to *insurance*, as it also does not do what the word promises. The damage is not really compensated, because the dead birds and fish remain just as dead after the payment, and most fishermen are probably just as bankrupt and sad as right after the disaster.

For this compensation, the tanker owner then gets a loan from a bank, just for *safety*. Next, the loan is insured by another bank because money is exchanged for risks, and from that moment on there are *CDSs*. Then the narrative is almost over. But not quite yet, because then someone thinks that such CDSs can just as well be *naked*, after which others enthusiastically buy a lot of those undressed things: things of which everyone has forgotten that they do not really exist because they are fiction based on fiction based on… and so on.

And then finally some people come and prick through that whole fairy tale, shouting: But hey, *money, borrowing, buying, insuring, CDSs*: they are all just words, not things! And suddenly everyone stops believing in them, and it is 2008 or maybe 2028, and companies go bankrupt, people are put out of the houses they thought they were the owners of, and some of those people jump from a bridge while others start living under them.

And then everything starts all over again.

And so, words are sometimes whole narratives. Some of these words are lies, such as *life insurance* and *compensation*. In any case, all *italicized* words in the above story are not real things but fairy tales, which are only believed by everyone because everyone believes them. Fun exercise: make *all* fairy tale words in the narrative italic. Then delete all italic words and see what you are left with.

There is a word for the whole collection of these kinds of fairy tale words. *Economy*. Obviously, this word is also a narrative, even a very complicated narrative. Economy is a fairy tale in which the whole world believes, at least everyone I know or see on TV. In fact, the fairy tale has become dominating the whole world.

So, this is what people do: they come up with words in order to tell narratives. To such narratives we then attach a new word, so that we can tell more narratives with them, and so we stack narrative upon narrative until nobody really understands it all, not even the 'experts' who have come up with the narratives. Because one expert makes up something crazy – let's say a CDS – and the next expert does something that the first expert did not intend: undress them, for example.

This is why the CDSs failed when they did not bring certainty but danger: the experts really do not know what they are doing.

All over again from the beginning

What did the economic and financial experts advise, and what did the governments do to prevent such a thing from happening again? They announced *laws and regulations*.

The European Union prohibited the trade in naked CDSs in 2012, well after the start of the Great Recession of 2008 for which the *housing crisis* struck the gunpowder in 2007, and linked stricter rules to other forms of investment.[2] That was a typical case of the pit that was filled after the calf had drowned: the damage had already been done.

The International Monetary Fund (IMF), however, found the EU measure nonsense: it would be unnecessary and perhaps even cause greater instability.[3]

Now, there are many leading economic experts working for the IMF, just as for the EU. Both powerful supranational organizations also receive a continuous stream of advice from leading professors from renowned universities and think tanks. The diametrically opposite opinions of the two organizations show wonderfully that it was indeed *that* and nothing more, what they proclaimed: *opinions*. No facts, no peer-reviewed science, no solidly substantiated and (therefore) generally accepted conclusions. Only opinions.

Another such combination of 180° opposing views arose – not for the first time – about how governments could best respond to the crisis with their government spending. *Cut back!* according to the more dominant economists. No, no: spend *more*, to stimulate the markets! others stated.

Once again, the financial-economic experts did not have a grip on what was happening.

Do you know of what level such a battle is? You may compare it with physicists who, while they figure out how to best bring a human being to the moon, argue with each other about whether you should leave *upwards* or *downwards*. If physics were still at such a stage, we would laugh at the physicists and send them to the prairie tarred and feathered. But the economists are believed, admired and rewarded with high salaries and with Nobel Prizes (as you will soon see). In short, physics has been a science for several centuries, but economics is yet far from there.

Moreover, it quickly became clear that the opinions of these experts were not even relevant. Forbid or not, those naked CDSs? Ha, the financial traders had already invented new derivatives in the meantime. We are now pleased with the trade in BTOs, *Bespoke Tranche Opportunities*.[4] They are similar to CDSs, but they differ because the composition of the 'basket' with which the BTO is filled is 'bespoke'; that is to say, it is tailored to the wishes of the purchaser. This is very useful for dealers in derivatives. Because, should the purchased BTO prove to be unreliable after all, well, then this time this is obviously due to the customer's own choices…

Oh yeah, and: the *subprime* mortgages returned again, less than ten years after the housing crisis of 2007, but that doesn't stand out, because there is a new word for them: they are now called '*non-prime*' mortgages. In the meantime, rules that have been drawn up in all sorts of countries to combat over-optimistic mortgages and derivatives are being abolished or relaxed. House prices rise again as flares, the next housing bubble swells.

How long will it take before the financial tower collapses again thundering? Shorter than you think, because: it has already happened. On a smaller scale, admittedly. Several times even, which is promoted because the commercial transactions for which people once needed hours are now carried out by computers in a split second. This gigantic speed means that there are occasional huge shock waves within a few minutes, where the prices of some or even all of the shares can suddenly collapse dramatically. This happened, for example, during the 'Flash Crash' on May 6 2010 at 2.32 p.m. East American time.[5]

You do not have to be surprised by the new crashes and threats. Economic crises are occurring constantly: much more often than most people know. That's what researchers Reinhart & Rogoff have put together for us, in their fascinating book *This time is different* from 2009. Their conclusions can be found in Figure 2.3.

The ironic book title *This time is different* refers to the reactions of people who suspect that we have now become so sensible and learned that economic crises never have to break out again. Do such people exist? Definitely, and they are not the lesser economic experts.

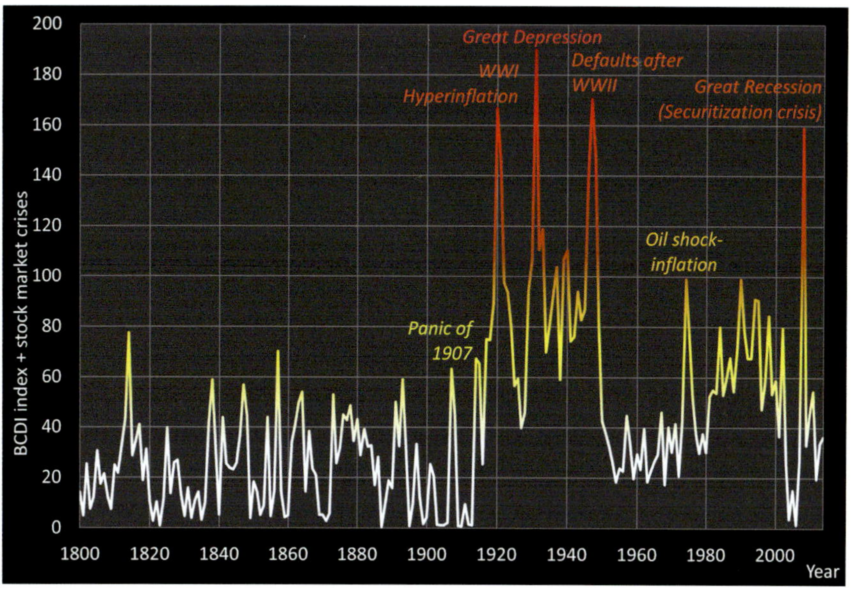

FIGURE 2.3 Economic crises between 1800 and 2014, according to Reinhart & Rogoff (2009, data updated 2018).

Take, for instance, the economist Robert Lucas, who won the Nobel Prize in Economics in 1995. In 2003, a few years before the Great Recession, he wrote[6]: "The central problem of the prevention of [financial-economic] depression has been solved, for all practical purposes." A year later, he received approval from Ben Bernanke, who was President of the Federal Reserve ('The Fed'), the US Central Bank, from 2006 to 2014. Bernanke expressed his confidence in the *Great Moderation*, the extinguishing (apparently, in the previous decades) of the economic cycles that lead to collapse at their lowest point. "Recessions have become less frequent and milder," he said in 2004,[7] and he added with satisfaction that "improved monetary policy" has "made a significant contribution". "This conclusion on my part makes me optimistic for the future." The housing bubble was already swollen ugly at that moment. The catastrophe began three years later.

Do you want more of such remarkable sayings? Here is Jean-Philippe Cotis, the chief economist of the authoritative Organization for Economic Co-operation and Development (the OECD) (the 'club of rich countries'). Standing on the edge of the volcano, he wrote in 2007, at a moment when, thanks to the American housing crisis, the recession had already begun: "The current economic situation is in many ways better than what we have experienced in years. (…) Our central prognosis is indeed very positive."[8]

But wait: it can be even more naïve. Well after the crisis hit hard and thousands of people had looked for bridges, another top economist was able to write that the economic crisis had not occurred, as economic crises *cannot* take place at all. This was Eugene Fama, who proved this interesting proposition on the basis of his *Efficient Markets Hypothesis* (EMH) of 1970: bubbles, like those of house prices, cannot occur at all, since the free market automatically makes sure that the prices are at the right level at all times, according to Fama.[9] For this brilliant statement, in 2013 he also received the Nobel Prize in Economics.* (As an aside: the word '*automatically*' will soon claim an interesting role in this book …)

I am infinitely grateful to all these economic leaders, since they prove, beyond doubt, that ordinary derivatives designers, average stock traders and directors of simple commercial banks are not the only ones who really do not understand economics.

2.3 Laws and regulations

The authorities tried to prevent new crises by promulgating laws and regulations, I just told you. They have been doing this for centuries, and it still does not help, because the crises keep coming. That is no wonder, because history, as illustrated in the graph of Figure 2.3, shows that the economy is incredibly unstable: every few years there are larger or smaller derailments. The situation can be compared to

* Fama's claim has been called "one of the most remarkable errors in the history of economic thought" (Shiller, 1984).

a giant snowball that bounces nervously up and down in a shallow hole on the top of a steep mountain, as in Figure 2.4. Inevitably, every now and then this ball rolls down with deep thunder, into the depths.

This small dimple on the mountain top, an island of stability in the middle of an ocean of instability, primarily concerns the *financial economy* (the trade in fairy tales like derivatives), and as such that does not matter much, of course, because you cannot eat money or build houses with it. But such a financial trap unfortunately also drags down the *real economy*, the trade in real things like potatoes and bridges, and that is pretty annoying.

And so, governments and international organizations come up with laws and regulations. They are preferably laid down in international treaties. The result looks like Figure 2.5.

I don't know if this kind of imposing measures – fencing and prohibitions – will impress you just as much as me. Keep in mind that traders and bankers are constantly trying to pull and hide all those fences and traffic signs because they find

FIGURE 2.4 Instability of the financial and hence also of the real economy.

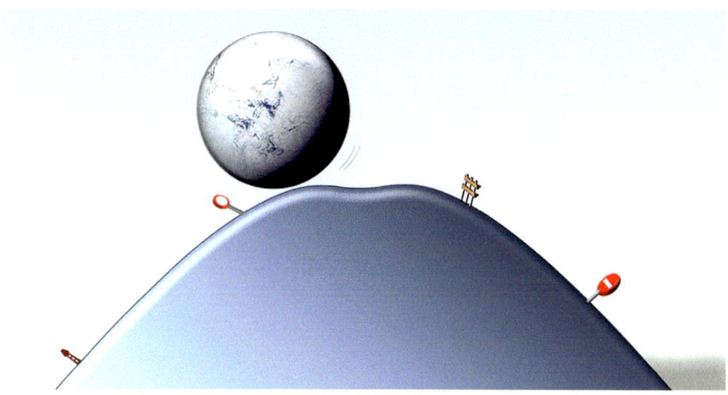

FIGURE 2.5 Attempts to combat instability using laws and regulations.

them irritating. Many of the rules that were adopted after the 2008 crisis – for example, in relation to sky-high bonuses – have already been abolished or softened.

This kind of policy does little or nothing. At best, they postpone the next crisis a bit, but experience shows that this will then probably erupt more heavily, because the bubbles and the illusions have more time to swell. Such policies are not effective, because they do not change the *fundamentals of the system*. The structures, the connections, the extremely complicated mutual forces and influences within the global financial-economic system simply remain the same. Worse: the way of thinking does not change. The narratives don't change, because the words with which these narratives are told remain the same. In other words:

Policy measures, such as the promulgation of laws and regulations, don't go deep enough. We must get to the deepest causes. We have to dig for the roots.

Recommended further reading (for details, consult the online Bibliography)

Book: Reinhart, C.M. & K.S. Rogoff (2009): *This time is different. Eight centuries of financial folly.*

Journal Article: Lin, Tom C.W. (2017): *The New Market Manipulation.*

3

FOUR SPHERES OF THOUGHTS AND ACTIONS

3.1 Words, narratives, policies, effects

What then does go deep enough? This is what I will examine in this chapter. But first I have to expand the terrain to its full extent. The instability of the financial economy is just one example of unsustainability that leads to catastrophes. It is an example from the angle of *profit*, one of the three main aspects of sustainable development. Together with *people* and *planet*, they are called the *Triple P*, a term we use after Ismail Serageldin who invented it in 1996, and John Elkington who made it famous.[10] Within the domains of the other two P's, there are also extremely dangerous causes of catastrophes.

Let me first indicate what exactly the three P's are about. They are also called *ecological*, *social* and *economic* sustainability. Together they deal with many topics, and especially about the complex relationships between all those topics. In business, the tripartition is also referred to as the *Triple Bottomline*. I offer you the explanations that I wrote for an earlier book.[11]

Triple P

Ecological sustainability (*'planet'*) relates to conservation and resilience of the natural environment. This implies that ecosystems and biodiversity are protected and that the ability of the natural environment to provide us with resources and regenerate our waste is not harmed.

DOI: 10.4324/9781003306269-4

Social sustainability (*'people'*) at an *individual* level is about respect for human rights, freedom and safety, cultural values, education and health, personal development, diversity, empowerment* and participation. At a *societal* level, it concerns peace, democracy, solidarity, and social cohesion.

Economic sustainability (*'profit'*) is present if development toward social and ecological sustainability can take place in a sufficiently stable economic environment and is financially feasible, and if individuals, families, and communities are guaranteed to be free of poverty. (Sometimes, instead of 'profit', the broader concept of 'prosperity' is used.)

Balance: Sustainable development means that all aspects and themes above are seen as mutually dependent and interrelated, in that the various interests, problems, and solutions are constantly and harmoniously weighed against and connected with each other. This principle is often referred to as: 'The three P's must be in balance'.

Now, in the previous section I have taken one instability in the *profit* zone – collapsing stock markets – as an example to show that we cannot combat the fundamental causes of unsustainability with superficial policy measures. You cannot eliminate structural defects with a layer of varnish.

You can also find such dangerous instabilities related to *planet* and *people*. Climate change, for example, can throw us into the abyss worldwide, as do collapsing ecosystems, an exploding economic growth, waves of dying plant and animal species (the '*sixth extinction*'), refugee flows of hundreds of millions of people, and torn societies. These are just six examples, two for every 'P', that I use in Figure 3.1 to illustrate that the range of structural instabilities can throw us into cliffs in many directions. I can easily add many more realistic threats to the figure, unfortunately. This will be done in Chapter 10.

And so, policy measures, however important and effective, can never be sufficient to bring us steadily into safer waters. *It has to go much deeper. More thorough. More fundamental.*

Words

It needs to be more fundamental. But how? By looking at the thoughts behind the policies being pursued. To the beliefs, the discussions, the decisions, the opinions, the principles from which the policies emerge. By looking at what the policymakers say and think; at the sentences they formulate; at the words that make up the sentences.

* Empowerment = 'the capacity of individuals, groups and/or communities to take control of their circumstances, exercise power and achieve their own goals, and the process by which, individually and collectively, they are able to help themselves and others to maximize the quality of their lives' (Adams, 2008).

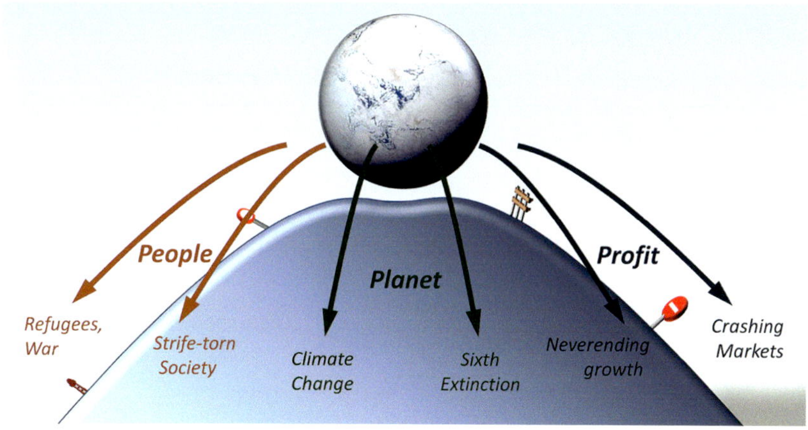

FIGURE 3.1 Abysses in many directions.

The words: they are the fundamental building blocks by which our thoughts, our narratives, opinions and choices are constructed. The use of words is crucial, because the words at our disposal determine what we can think, say, and decide. Words steer.

As a first step, therefore, I describe the consequences of the choice of the words with which we tell our narratives. You may already feel it coming: in order to tell better narratives, we have to use different words. Because:

> *In order to say something, you need* ***words****.*
>
> *Those who have words, can tell* ***narratives****.*
>
> *Anyone who tells narratives, can make* ***decisions*** *and thus choose* ***policies****.*
>
> *Whoever chooses policies, determines the* ***events*** *and therefore the* ***effects*** *thereof.*

This means: the words we have are decisive for what happens and what the consequences are. Words determine how the world looks after a while. You can put that in a picture like Figure 3.2.

This figure consists of four layers, which I like to call 'spheres'. I have given them decent names, derived from Greek: from bottom to top I call them *logosphere*, *mythosphere*, *politosphere* and *phenosphere*.*

With the aid of the image of the four spheres you can discover a couple of beautiful truths:

> *The words that you possess determine what you can say.*
>
> *The words that you* ***don't*** *possess determine what you* ***can't*** *say.*

* *Logos* = word. *Mythos* = narrative. *Politike* = policy. *Pheno* (*phainein*) = reveal, as in phenomenon and phenotype.

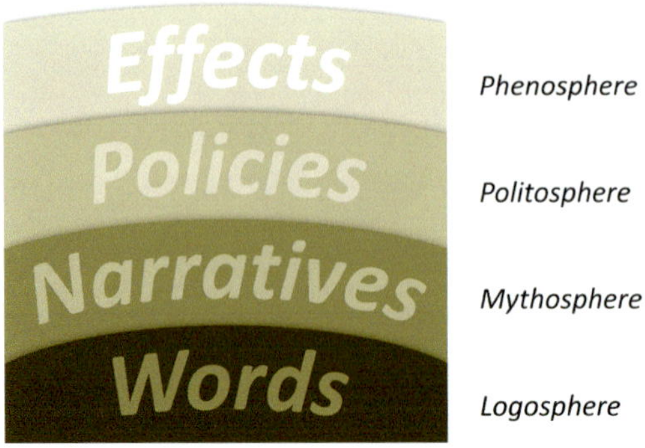

FIGURE 3.2 The Four Spheres of human thinking and acting.

You can try out this simple principle for yourself. First, pretend you have no words for colors and smells, and then describe an autumn forest.

Or: how would a king who has no words for numbers over 4 handle taxation?*

And how does a society that doesn't know the word 'human rights' treat its citizens (and slaves)?

A final one. Complete this sentence: Emperors and presidents who don't know the word 'peace' will …

But there is another, even more important conclusion. The image of Figure 3.2 also shows:

> *If you possess flawed words, or use words in a flawed way, you will tell wrong narratives, which means that you conduct bad policies, causing harmful effects.*

This is the deepest cause of all unsustainability in the world. I will illustrate this by means of an example.

Flawed word use: an example

The word '*value*' is a very flawed word. Perhaps the word is fundamentally wrong and we should abolish it. But it is also possible that the word itself is quite okay but that some people use it in a rather wrong way. Unfortunately, those people are mostly those who determine for a large part what is happening in the world, as quite a few economists and bankers, together with members of Accounting Offices

* Mathematical historian Georges Ifrah (1981) gives fascinating examples of this. For example, the members of the Aranda tribe in Australia only knew the numbers one, two, two-and-one, two-and-two, many. Also, the Indians of Fire Island, the Abipones in Paraguay, the Bushmen and Pygmies in Africa and the Botocoudos in Brazil counted no further than 4.

and governments, think that '*value*' is equal to '*money*'. In Chapter 7 (page 78) I will call this the 'Value Myth'.

This is not true, of course, and you and I know that very well. 'Money' is a fairy tale, as I mentioned, while 'value' is – well, actually the most important thing that exists. Check this out: if you list the most valuable things in your life, there is a good chance that you will put things into your personal Top Ten such as:

- *The love of and for your partner, your children, your parents*
- *Health*
- *Your faith or your belief*
- *A smile*
- *Freedom, peace, friendship, trust, respect, appreciation, recognition*
- *The beauty of a sunset, the silence and serenity of a forest*
- *Life fulfillment, self-realization*

The striking thing is that, among the things that are most valuable, *none of them can be bought with money*.

Quite a few of the most influential policymakers in the world don't understand that. And thus, you get silly calculations.

Numerical example: one racehorse

1 racehorse	=	*10,000 chickens*
2 loafs of bread	=	*1 beer in a café*
1 Mercedes	=	*250 bicycles*
1 semi-automatic rifle + 1000 bullets	=	*2 funerals*
1 Time Warner Company	=	*2440 forests of 3400 hectares*
1 CEO	=	*6500 employees on the minimum wage*
1875 mice	=	*1 donkey*
10 euros	=	*12 dollars*
2.4 grams of gold	=	*1 dental restoration (deep filling)*
2 children	=	*97 holidays to Thailand (2 persons, 3 weeks)*
8,000 times fun together at the theater	=	*1 cure for a cancer patient*
1 Google, Inc.	=	*200 billion eggs*
9 million shirts made by child slaves	=	*1 footballer*
1 economic crisis around 2008	=	*22,000,000,000,000 dollars*
1 human (US)	=	*588 humans (Central Africa)*

The equations are really true. Financially speaking. (Or they were in the recent past, factually or on the basis of expert estimates.) If you want to know why and how, you can download a spreadsheet ('*What's it worth*') from the website associated with this book, in which the value of each component is shown in dollars and euros, complete with explanations and source references.*

Even though the comparisons are true, many conclusions are clearly absurd. For example, the purchase costs that customers pay in shops in Western countries for nine million shirts may demonstrably be equal to those of the transfer of one footballer (Neymar to Paris Saint-Germain in 2017 for €220 million), but no one is able to feel the suffering of tens of thousands of children in India, Bangladesh or elsewhere who have stitched the shirts together;[12] or to estimate the value of the opportunities they miss out on a decent life because they can't go to school. Is this all really equal in value to a few extra goals that PSG might score? The suffering and missed opportunities for a decent existence of the children sold by their parents (in order to enable them to feed their other children) can never be expressed in money.

No, not everything of value can be expressed in money. When you try to position the smile of a loved one somewhere in the calculation, is it doubtful whether you will succeed. The same applies to the smell of fresh bread, for a memory from your childhood and for your newborn child. Certainly: the example includes the price of '2 children', which is equal to 97 vacations. But that concerns an estimate of the average amount that you will spend on food, clothing and housing, as a result of your choice to have children: approximately US$230,000 per child in the period between birth and majority – at least for an American middle-class family, according to the US Department of Agriculture.[13] The happiness and pleasure that you experience through that choice, but also the worries and sorrows that are likely to come your way, cannot be expressed in money. This certainly also applies to the happiness and grief that your children will experience as a result of your decision to let them exist.

The logosphere

Nature is exhausted by overexploitation, overgrazing and overfishing. There are mass extinctions of both plants and animals. The planet is contaminated with toxic substances, particulate matter and greenhouse gases. This happens because people tell false narratives. Cut down a forest and sell the wood – says one of those narratives – because then you have a *profit*. 'Profit' is a wrong word (or, perhaps, a good word that is used incorrectly).

Bury your waste in the ground. Discharge waste water and plastics in the sea and spray CO_2 into the air, and they all nicely disappear! Says a different narrative, which tells us that the ground, the sea and the air are *externalities*, which we can freely dispose of: they don't count, they don't matter. 'Externalities' is a wrong word.

* I wrote it in chapter 1 already: the link to download tools like this can be found at https://niko.roorda.nu/books/omniconomics.

FIGURE 3.3 The Four Spheres, filled in with traditional words.

Take *ownership* of the metals in the ground and do whatever you want with them, because if future generations ever want to do that, well, we don't hear or see them. 'Ownership' is at least a strange word. A wrong word?

The narratives that are told with these kinds of words are wrong, because they lead to policy choices and behavior that evoke disastrous effects. I have inserted these words, along with a few others, in Figure 3.3 in the logosphere. Of course, I don't offer a complete list of such words, that would not be possible at all. If you want, you can add words yourself, as many as you want. Obviously not all words in the logosphere of Figure 3.3 are wrong (fundamentally or in the way they are used). If we want to find out which words are okay and which ones are not, this yields a fascinating search. This will happen soon.

The mythosphere

In the second layer from below, the mythosphere, I have filled in a number of narratives. Of course: those narratives are also indicated with words. That is what we do after all: we choose words with which we tell narratives, after which we also give a word to a narrative as a whole, so that we can stack new narratives on top of the old ones. *Capitalism*, for example, is a narrative, stacked on words such as *value*, *profit*, *enterprise* and *property*.

But how do you know now which words belong in the bottom sphere and which ones in the higher spheres? My answer is: the logosphere contains the words that are the most fundamental in our thinking. The basic concepts, the key words that shape our thinking. The *paradigms*. The *memes*.* In the second sphere you then find the words that have a more complex nature: the narratives. Not only beautiful little narratives such as good jokes and fairy tales from Grimm, but also – above all – the Big Narratives, which have shaped the history of mankind. Many narratives I have filled in are well known to you; others, including *automatism*, I will explain below. Also, 'economy' is a narrative, which tells us that it is possible and useful to set up and manage the world according to economic principles in the logosphere.

I admit immediately: the division between the first and the second floor is, to a certain extent, subjective. That's why you can consider the details of Figure 3.3 as a proposal. Please feel free to distribute the words differently among the boxes, to remove words and to add other words. Or even to insert extra layers: spheres that are located between the ones I have designed. I claim no accuracy and also no completeness. Make your own schedule, if you think mine is wrong!

The politosphere

In the third sphere, the politosphere, you find the choices people make and the routes they follow, based on the narratives in which they believe. For example, for whom *profit* is a basic word in the logosphere, *profit-making* is a goal and policy choice in the politosphere that indicates the direction of concrete behavior.

The phenosphere

The fourth sphere, the phenosphere, deals with the consequences of the choices and behavior of people. Some of those consequences are far from wrong. *Prosperity* is an example. And *quality years*: a growing number of people live more and more years in good health and wellbeing, i.e., in good quality.† Unfortunately, the phenosphere also contains many negative effects that are extremely harmful and dangerous and therefore even threaten the beneficial effects.

3.2 Interventions in the Four Spheres

With the help of the Four Spheres, I want to investigate what people tend to do, in order to deal with unsustainability issues. By way of illustration first: this.

* Meme = unit of cultural transfer: An element of a culture or system of behavior passed from one individual to another by imitation or other non-genetic means (Oxford Dictionary).

† In full: Quality-adjusted life year (QALY), a concept introduced in 1968 by Herbert E. Klarman et al.

On October 3 1997, tailor Zekeriya Gümüş and his family were expelled from the Netherlands. It happened against the will of millions of Dutch people; in spite of protests by thousands of Amsterdam residents; in spite of resistance by hundreds of politicians, lawyers and media makers; in spite of a motion supported by almost seventy members of the Dutch Parliament; and after at least ten unsuccessful court cases.

Gümüş had come to the Netherlands illegally in 1989, followed a few months later by his wife Esme and their 6-year-old son Ramazan. But Gümüş and his family came from Turkey, which is a safe country, according to the guidelines. Gümüş was one of the 'economic refugees', which is better not to be: they are "only out to get a piece of our wealth". People from unsafe countries such as Iraq or Afghanistan may be granted 'political refugee' status upon arrival, requesting asylum and perhaps, in time, a permanent residence permit, as their lives are in danger.

But what is safe? Ramazan's younger brother Samet died in Turkey as a six-month-old baby from malnutrition and poor medical care. And whether you are killed by a hostile regime in Iraq, or by starvation in Turkey, you are just as dead.

Gümüş worked hard: first as an employee in garment workshops, later in his own business. What he didn't know was that his first employer didn't pay any tax for him, so he didn't work 'white' for a few years. It was precisely during those years that a regulation was passed, stating that illegal foreigners could be legalized if they had worked 'white' in the Netherlands for sufficient years. Unfortunately, Gümüş did not meet that requirement. After several years of legal battles, mass demonstrations in the streets, attention from newsreels and newspapers at home and abroad, and direct interference from a mayor and the prime minister, the Gümüş family was put on a plane to Turkey in October 1997. Including Ramazan, 14 years old, who had already attended Dutch schools for eight years, and his little brother named Samet (like the deceased baby), who had been born in the Netherlands and had never been to Turkey.

This distressing, and in many people's eyes unacceptable, outcome is a typical example in which the four spheres can be clearly recognized. From top to bottom:

Phenosphere: the *EFFECT* was that a totally Dutchified family had to leave because it did not comply with the rules. The Secretary of State and the Prime Minister concerned could have intervened at the phenosphere level by making an exception, but they did not: "Rules are rules", they explained.

Politosphere: To avoid the undesirable outcome, but, in accordance with the rules, the *POLICY* would have to be changed. However, that had just been painstakingly laid down in the new rules a few years earlier. A change was politically unfeasible and would also have taken too long.

Mythosphere: All this could only happen because we believed in the *NARRATIVE* called 'country', formally: 'nation-state'. That narrative is told in words like borders, citizens and passports. In Europe, people also tell a very different narrative, called 'the European Union', which is translated in the politosphere as 'the Schengen Agreement', with the effect in the phenosphere of free movement of people. But then again, Turkey is not an EU member and outside the Schengen zone.

Logosphere: The way of thinking, anchored in countries, borders and expulsions, relies on the *WORD* 'illegal', which means something like 'does not belong here'. It is a dangerous word, because although it actually refers to 'illegal residence', it is widely understood to mean 'illegal person', so that 'should not be here' becomes 'should not be'. As a result, the mere existence of a person is deemed illegal and therefore criminal, thus laying a firm fundament for discrimination.

Another dangerous word is 'economic' as in 'economic refugee': someone who is such a refugee, as opposed to a genuine 'political refugee', should not be faint-hearted; they should just go back to their own country and lead a miserable life there – possibly losing a child or two to hunger. And so, you see concretely how the words in which we think determine our narratives and thus subsequently our policies and their effects.

After this illustration, I will again go through the four spheres, but this time in a more general way, as shown in Figure 3.4; starting again at the top: in the phenosphere.

Interventions in the phenosphere: symptom control (time perspective: right now)

When people focus their actions on the phenosphere, they are not concerned with tackling the causes of problems – which belongs to the politosphere and below – but with symptom control or with learning to live with the consequences. In a country such as the Netherlands, you may, for instance, increase the sea and river dykes, now that climate change causes a sea level rise and a whimsical, occasionally heavy rainfall. Against dangers elsewhere in the world, people can build bunker-like dwellings, resistant to earthquakes and severe storms. A special kind of bunker is the *Svalbard Global Seed Vault*[14] – nicknamed 'Noah's Ark' – on the icy island of Spitsbergen. There, as many seeds of plant varieties as possible are stored, including many food crops.

In countries where things are going well, residents can pick up refugees from areas where things got worse. If calamity then strikes in the once-safe havens, it will be time for the wealthy too to flee destitute. You see all these options in Figure 3.4, in which, for the sake of convenience, I swept to the left the words shown in the previous image. In the new figure, in the phenosphere, you can see some other choices you could make. A much-chosen option is 'denying', for example

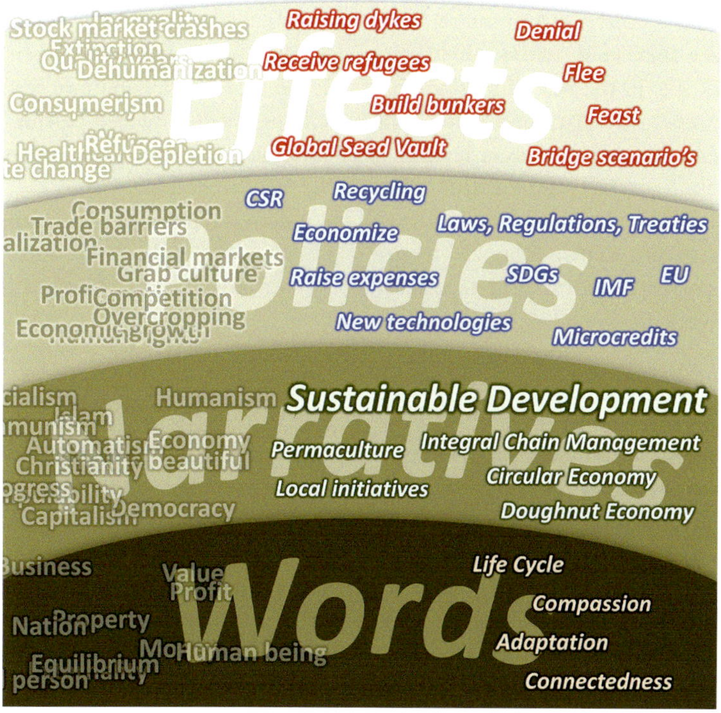

FIGURE 3.4 The Four Spheres, supplemented with modern elements.

climate change – presumably usually against better judgment. This choice of life is often combined with 'partying' until the world decays, although a different sort of end may be reached even before, through one of the two bridge scenarios I mentioned earlier. Both belong to typical fight-, flight- or freeze-behavior, often combined with behavior like getting angry with the messengers – the scientists and the press –, running away, or closing eyes tightly.

Interventions in the politosphere: repairs (short-term policy)

If you next look at the politosphere, one sphere deeper, then Figure 3.4 helps to see why laws and regulations do not go deep enough. They are, just like budget cuts or additional expenses, examples of short-term policy. They are useful, but they do not change anything about the two spheres below. You can introduce or abolish laws, as many as you want. You can set up international organizations, such as the European Union, the United Nations, the International Monetary Fund, you name it, and that's great. You can agree on worldwide *Sustainable Development Goals* (SDGs). They certainly result in important successes. On a smaller scale, of industrial branches or individual companies, you can work on recycling, on CSR, on new technologies.

Individual people can also make a significant contribution. In our private lives, in our role as citizen and consumer, we can separate waste, use energy economically and place solar cells, refuse unnecessary packaging and hundreds of other things. At work, i.e. in the role of a professional, we can also all contribute to sustainable development, in whatever profession or branch; as a self-employed person, or as a manager or employee at every level in an organization. I illustrated this in several of my books[15] on the basis of a large collection of real-life examples.

In all these ways, the damage caused by the system is repaired. Do you remember how I started the current book? "It's great," I wrote. "No, really. The efforts people spend on sustainability." But repairing the system is *not enough*. Because the Big Narratives have to be told again, too.

Interventions in the mythosphere: renovations (medium-term)

Fortunately, this happens too. In Figure 3.4, look at the mythosphere: at the narratives you see there. An important example is the narrative of the *circular economy*, which tries to bring the cycle of thought from nature into the heart of the economy. Common names for this narrative in the 1990s were *design for disassembly* (DfD) and *integrated chain management* (also known as *Supply Chain Management*, SCM). Later, a distinguished circularity ambassador, Michael Braungart, named the narrative C2C, *cradle to cradle*. The next top ambassador was Kate Raworth, who named it *doughnut economy*.[16] Although the narratives are retold repeatedly, becoming increasingly stronger and more comprehensive, the basic idea remains the same: the closing of cycles, just as nature has already done for a few billion years. You can find the basic idea of the circular economy, with its characteristic two cycles, in Figure 3.5.

Another narrative in the mythosphere is the counterpart of the massive globalization trend: '*small is beautiful*', with a plea for local markets and communities. The name of the narrative is taken from the title of a famous book by Ernst Friedrich ('E.F.') Schumacher from 1973. New narratives in the same theme are told in the 21st century, in the form of hundreds of small-scale initiatives, for example with local currencies such as the *Chiemgauer* in Austria, the *natas* in the Dutch town of Emmen, the *blés* ('*Bon Local pour l'Économie Solidaire*') in Grez-Doiceau in Belgian Wallonia, and the *Salt Spring Dollars* in Canada. You can also include *permaculture*, as well as all kinds of organic farming.

All these new narratives, from circular economy to permaculture, are part of one wonderful narrative that encompasses them all. This is *sustainable development*: the Big Narrative of the 21st century. This is much more than carrying out repairs in the politosphere. Interventions in the mythosphere are a major renovation of the global economic system.

Interventions in the logosphere: from the ground up again (long-term strategy)

They are beautiful new narratives, out there in the mythosphere. But there is a major limitation. They largely continue to *use the same words*. Circular economy still

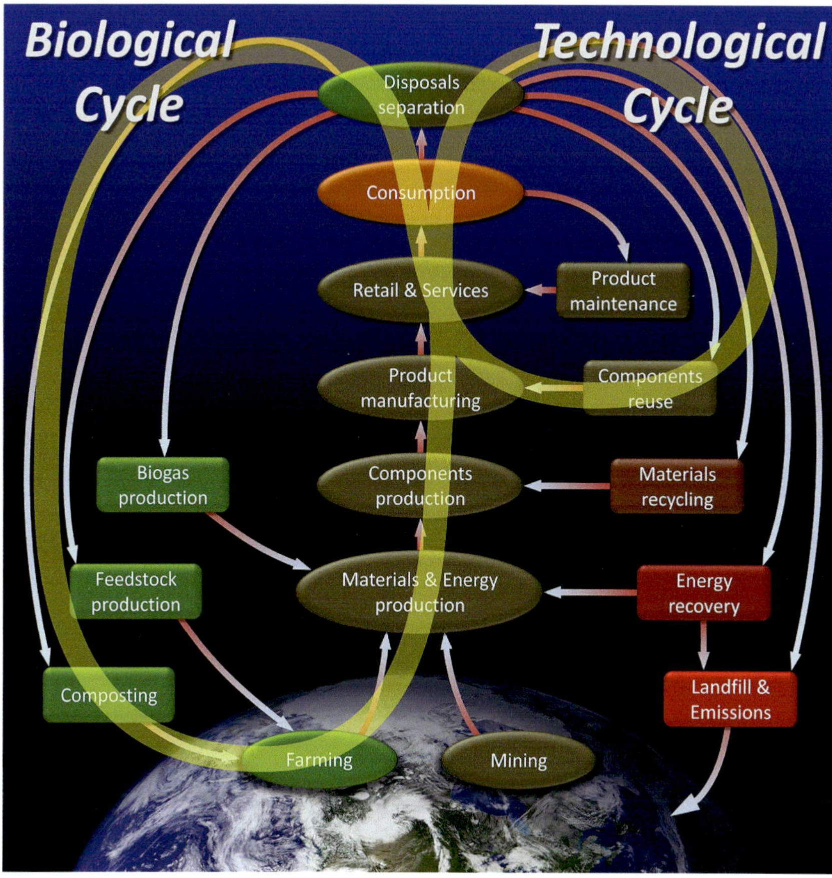

FIGURE 3.5 Circular economy, with its characteristic two cycles.

talks about profit, value and ownership. About legal entities, companies and states. About money. This even applies to 'small is beautiful', although the supporters do not seem to want to recognize that. For example, remember that local coins are also just *money*.

Telling new narratives with old words: that is not enough. Because those words are not correct, as I am going to show.

It is true: new words have indeed been added. Just look at the bottom row in Figure 3.4, in the logosphere. I have added a number of words that are strongly related to circular economy ('life cycle'), to 'small is beautiful' or to other aspects of sustainable development. It is these words that make it possible to tell a Big Narrative about sustainability right now. More and more people tell this narrative. I am one of them, telling the narrative enthusiastically since 1991. Over the years, the narrative is becoming stronger and more comprehensive, and that is wonderful. A recent addition to the narrative are the SDGs I just mentioned: the so-called '*Agenda 2030*', in which concrete goals and policies have been agreed internationally for the period from 2015 to 2030.

And yet all that is not enough. Because the interventions in the *phenosphere* are ultimately nothing more than symptom control – or not even that. Policy-based interventions in the *politosphere* are short-term approaches, which is true even for the mythospheric, imposing and reasonably successful SDGs, because they have been defined for a period of only 15 years, and although they do improve conditions, they do not remove any deep causes.

Recommended further reading (for details, consult the online Bibliography)

Book: Elkington, John (1998): *Cannibals with Forks – The Triple Bottomline of 21st Century business.*

Book: Roorda, Niko & Anouchka Rachelson (2018): *The Seven Competences of the Sustainable Professional. Developing best Practices in a Work Setting.*

Book: Raworth, Kate (2017): *Doughnut Economics. Seven Ways to Think like a 21st-Century Economist.*

Book chapter: Asdal, Åsmund (2021): *The role of the Svalbard Global Seed Vault in Preserving Crop Genetic Diversity.*

4

INTRINSIC SUSTAINABILITY

4.1 Glued-on versus intrinsic sustainability

The new Big Narrative added to the mythosphere, sustainable development, offers us a new way of looking at reality. This is excellent for the medium term – roughly one or two generations. But the narrative is *added* to the existing narratives about human society. 'Glued-on'. The sustainability narrative is thus not yet integrated into the heart of the global human system, as it does not change the fundamental structures.

That is why here is the core of my message in this book:

> **Intrinsic sustainability**
>
> *If the narrative about sustainability as it is now told will be successful, the world will become a lot more sustainable. But: humanity will have to continue to work hard for years, centuries and millennia to keep the world sustainable. And then once, sooner or later, there comes a time when the sustainability effort weakens, after which society slips in the direction of unsustainable catastrophes.*
>
> *That is why the ultimate task of the sustainability narrative is, to create a world that is automatically, autonomously, attracted towards sustainability, without people having to make an effort.*
>
> *This is the principle of intrinsic sustainability.*

The successful execution of the assignment to bring about intrinsic sustainability will provide us with the ultimate long-term solution. I do not know what that solution will look like. I cannot tell you, because *nobody* can do that right now. We do

DOI: 10.4324/9781003306269-5

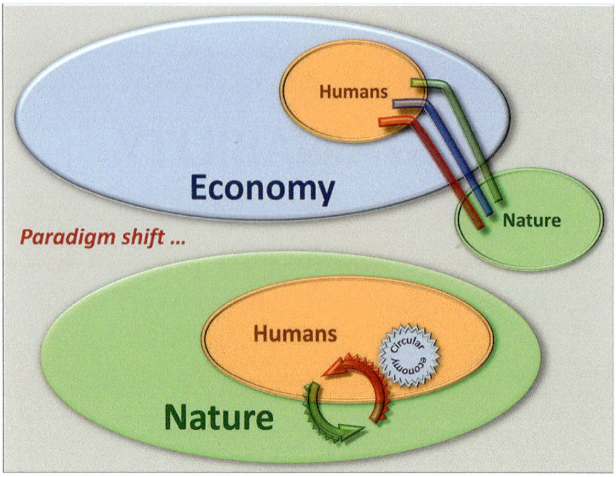

FIGURE 4.1 Paradigm shift: from economy to omniconomy.

not yet have the necessary knowledge and insights. In fact, the words we use today are not good enough for it. So, we have to invent new words.

What I do know is that we have to replace an existing narrative. The new narrative will be so drastically different from the old one, that it must get a new name. The old narrative, which has to be abolished, is called *economy*. I call the new narrative *omniconomy*. Figure 4.1 shows in a compact way why this paradigm shift is necessary.

What Figure 4.1 shows in the upper half is the current dominant view in economic thinking. According to that view, economic science (*economics*) describes the economic system (*economy*). Many actors are part of that system: companies, states, infrastructure, patents, and people. Economics regards people as economic things, which is expressed in a (for me) horrible way in the term '*human capital*': one of the terms that I think should be banned from the past as soon as possible, because it mocks human dignity.

Outside the economic system, according to the traditional view, there is nature, which is referred to by an equally horrific term: '*externality*'. The main task of the economy is to extract valuables from nature as efficiently as possible and to *internalize* them into the economy. Economists call this *value creation*. The economy is the 'universe': only what is in it, is important; everything else is irrelevant.

It is this dominant view – people are elements within the economy, nature is something external – which is the cause of many catastrophes. In the context of sustainability, however, a fundamentally different vision unfolds, which you see in the lower half of Figure 4.1. According to that view, our 'universe' consists of the entire planet Earth, that is, of the living and the non-living *nature*. People are part of nature, including everything we have produced: houses, highways, internet, money; and also love, loyalty, solidarity; as well as aggression, hatred, war.

In the new vision, people are not part of the economy; on the contrary: the economy is part of the human activities – besides many other activities that are not economic in nature (but often drawn into the economy by means of money): for example, art, music, dance, poetry, gameplaying, coziness, eroticism, relaxation, peace.

This implies that the economics discipline should not be focusing on an incomplete system which, as a part of the complete system – planet Earth, our 'universe', our 'home'[17] – is related to only an artificially limited set of properties, relationships and developments. An amputated reality. In certain situations, such a limitation can be useful. But when it comes to developing a long-term global policy, it is an unacceptable curtailment of the image: it provides an eye-blinkered vision at best.

As the so-called 'economic science' nevertheless tries to study the economy as an isolated system instead of as an aspect of a larger system, economic theorists use terms that are misleading: wrong words, such as *externality* and the humiliating *human capital*. It is therefore no wonder that the field of economics fails in all sorts of ways, resulting in looming catastrophes in each of the three P's. The discipline 'economics' is therefore not a science at all: it is a *protoscience*. I am going to tell you soon what that is, and next I will show that the economy is certainly not a mature science.

The same applies, by the way, to some other so-called 'sciences' that try to describe aspects of our 'universe'. You can think of political science. Of sociology and ecology. Each of these disciplines studies a (too) limited selection of all relevant variables and processes in our Earthly universe. For the realization of intrinsic sustainability, a new science is needed, which integrates all these areas, so that the various aspect systems do not act separately but within their complex context. I call this integrated science *omniconomics*: the prefix 'omni-' stands for 'everything', just like in 'omnibus' and 'omnivore'.

I don't know what this omniconomic science will look like. I don't know which words the new discipline will use or which narratives will be told with it. But I do have an idea about a route along which we can discover it together. I propose a plan.

4.2 The Broad Holloway

The plan is based on three preconditions that this plan must meet. I start with them, so that you know what to expect – and what not. I formulate the preconditions in the form of answers to three urgent questions.

First question: is this possible, intrinsic sustainability?

I know, it sounds pretty fantastic: sustainability that develops naturally. With a nice word: sustainability that is *system-integrated*. Sustainability will – such is my goal – become a basic feature of the global omniconomic system, which can no longer disappear or fade away all by itself.

So, the question is: is that too good to be true? Is intrinsic sustainability a fairy tale, a beautiful dream that will never come true?

My honest answer is: I cannot prove that intrinsic sustainability is feasible. On the other hand: you can't prove that it is not. This is a matter of principle: by definition it is impossible to prove that something is not feasible. (See Appendix 1 for a further explanation.) Although sometimes people try something like that: just look.

Space travel or air castle?

On December 30, 1959, Prof. Dr. George van den Bergh (1890–1966) was interviewed by Bert Garthoff in a Dutch television broadcast. The conversation was published as a book with the title: 'Space travel or air castle?'.[18] Here are a few quotes.

> *"My point of view is that space travel in the real sense of the word, that is to say as manned spaceflight, where one or more people leave the earth to another celestial body to stay there for some time and then return to earth, is not possible. Not now and never."*
>
> *"Space research has progressed significantly in a few years since the first Sputnik. It is already possible to confine a man in a cabin with all that goes with it and to give such a cabin with a so-called four-speed rocket the speed of more than 10 km per second, which is necessary to overcome the gravity of the earth. This is indeed possible, but there remains a fundamental issue: how does one get a human being alive again on earth? For no one will want to send a person into space if he knows that he will never return. Well – this 're-entry' problem is not resolved and, in my opinion, it is unsolvable."*
>
> *"Think of the smallest actual journey through space, i.e., to the moon. We approach the moon and we see it from our vehicle. Still 10,000 km to go, another 5000 km (we go fast), another 2000 km – it is a fast fall to the moon. We fall at a speed of thousands of kilometers per hour. Another 1000 km, 500 km – Fire! We are slowed down, but of course not completely. There is still some speed left and we fall further. Now there must be breaking speed again, weaker this time. So, it is theoretically possible that one finally comes down on the moon as soft as a rose petal on a quiet pond. On earth one can always catch that last speed with a parachute, but on the moon that is not possible, because there is no atmosphere. Even the biggest parachute does not help. One can only work with counter-fire and even if only a speed of ten meters per second is left, which is almost nothing in relation to the previous speeds, then one still crashes."*

On April 12 1961, Yuri Gagarin was the first human being to leave the planet Earth and to return to the ground after a little bit more than an hour. On July 20 1969, two crew members of the Apollo 11 mission landed on the Moon, which they left after about five hours. The Moon travelers returned safely and landed in the Pacific Ocean on July 24, after which they were brought home by the USS

Hornet. Manned trips to the planet Mars are being prepared by various parties for the years after 2020.

For ten thousand generations or more, most of our ancestors have believed that people would never be able to fly. Today, more than one million people are in the air at any time of the day (excluding the period of the Covid-19 pandemic). And although it has been 'proven' that space travel is not possible, space outside our planet has been permanently inhabited for a number of years now. Air castles *can* be built.

Do you know the *Three Laws* of the visionary science fiction author Arthur C. Clarke?[19]

Clarke's Three Laws

First Law: *When a distinguished, but elderly, scientist states that something is possible, he is almost certainly right. When he states that something is impossible, he is very probably wrong.*

Second Law: *The only way of discovering the limits of the possible is to venture a little way past them into the impossible.*

Third Law: *Any sufficiently advanced technology is indistinguishable from magic.*

So do not shout too quickly that something is not possible. Just try it, because that is the only way to discover if it is possible. As long as it fails, it can still succeed afterwards. So: keep trying, with My First Law in mind:

Roorda's First Law: *Only when it succeeds do you know whether it is possible.*

This leaves us with no other choice than to believe that intrinsic sustainability can be realized, to work on it with all our strength, and to make it happen. We're just going to build it, this intrinsic sustainability. Do you join?

Second question: does intrinsic sustainability imply that nothing can ever change once it's realized?

There are also people who, when hearing what I propose, are afraid that intrinsic sustainability should ultimately lead to a society in which, once it is realized, nothing may ever be changed. If at some point we have reached a kind of ideal balance point, we must remain anxiously there. Is that so, does *stably* sustainable also mean *statically* sustainable?

The answer is 'no, fortunately not'. Fortunately, because otherwise we would achieve the end of the human adventure with the realization of intrinsic sustainability. It would mean the 'end of history', to quote the title of a famous book.[20] What would life become boring then! The answer is 'no, it never becomes static', and I can show why, on the basis of something I once experienced as a little boy.

I remember vividly. I was terrified. My parents had taken my brothers and me for a day on the beach. For the first time in my life, I climbed a dune, arrived at the top, and decided to run down to the beach towards the sea as quickly as possible. I ran down, but what was happening? I could not stop anymore! I threatened to fall over, and to prevent that I had to run faster and faster! I have screamed with fear.

Fortunately, there was an unknown gentleman who was walking in front of me. He heard my screams, turned around and caught me. Saved! The sweat splashed from me on all sides.

In this *almost* painful way, I discovered an important principle that we need in order to stay on our feet. Much later, as an adult, I have put the principle into words. *When you move, move through a continuous series of equilibrium conditions.* That does not mean that you have to hold still at your place. You can move all right, even move quickly to new locations if you wish, but in such a way that you can immediately stop at any moment.

You can also compare it with someone walking down a staircase. You, for example. If you keep the center of gravity of your body above your feet at all times, then you know for sure that you are stably going down in a quiet, controlled manner. If you discover something you want to look at, halfway down the stairs, you can immediately stop without problems. While in motion, you are constantly in balance. If you do it differently, you run the risk that you have to walk harder and harder to avoid falling over.

Alas, this image is perfectly applicable to our current worldwide society. It does *not* move through equilibrium conditions. The technological, scientific, political and economic developments are going so fast that we are not in balance for even one moment, so we have to walk harder and harder to avoid falling over. It goes without saying that we cannot sustain that acceleration indefinitely. In this occasion, there is no friendly gentleman halfway down the dune that catches us (although some people may think otherwise).

That is why we have to achieve intrinsic sustainability. That does not mean that we have to freeze; the end of the human adventure is certainly not in sight. Intrinsic sustainability means that we discover how we can further develop safely: via a continuous series of equilibrium states.

I go one step further. If we, as humanity, go through new developments over and over, it goes without saying that we will always come up with new words based on

FIGURE 4.2 The 'Flywheel': a cyclical process in which effects evoke new words.

those experiences. With which we tell new narratives. Those words and narratives are fed by the effects of the narratives that we told earlier. Together, this means an infinite development, an always-ongoing iterative process. I call it the *'Flywheel*', which I show in Figure 4.2: words generate narratives, generating policies that generate effects that generate new words from which new narratives ... and so on, until the Sun or humankind extinguishes.

Third question: does intrinsic sustainability guarantee eternal stability?

Will intrinsic sustainability, if we know how to realize it, guarantee for eternity that everyone will find Paradise on Earth? Will society become permanently and guaranteed stable and everyone happy?

No, we cannot make it that beautiful. If only because there is always the risk that on an evil day a block of stone with a diameter of fifteen kilometers collides with the Earth, as happened repeatedly in the past. The last time it occurred about 66 million years ago, wiping out all dinosaurs together with many more animal and plant species. According to Arthur C. Clarke, our current technology may not be indistinguishable from magic, but protecting us from such a meteorite will far exceed our present magical powers.

There are more possible causes of major catastrophes that we can never fully prevent. Heavy earthquakes; volcanic eruptions obscuring the sky worldwide; pandemics that strike even harder than the Black Death, the Spanish flu, HIV/AIDS or Covid-19: it would be an illusion to think that we can ban them completely, because these are forces that are immensely greater than ours. Perfect stability, as symbolically shown in Figure 4.3 in the form of a snowball between two infinitely high mountain walls, will never exist.

FIGURE 4.3 Perfect stability between infinitely high mountain walls is not realistic.

FIGURE 4.4 Broad stability, however, is realistic.

In addition, there will always be things going wrong on a smaller scale. On a personal level: babies die, children are born with errors in their genetic material, car accidents happen and marriages break up. Our Earth will never become a perfect Paradise.

And so, intrinsic sustainability is not about perfect stability but about a significant increase of stability. Figure 4.4 shows this *broad stability* symbolically. Compare that with the narrow dimple on top of a steep mountain top in Figure 2.5, to which fences and traffic signs have been added. It is immediately clear that such powerless policy instruments become superfluous if the natural stability of the system is sufficiently broad.

Nevertheless, Figure 4.4 does not yet give the correct image, because this image seems to suggest that intrinsic sustainability leads to immutability, and such rigidity has been contradicted in response to the second question. That is why Figure 4.5 shows a better picture. Broad stability there is not a wide hole with one lowest point, but a path stretching into the future, beyond the horizon. Intrinsic sustainability is not a static but a dynamic stability.

That path is beautifully shown in Figure 4.6 in the form of a photograph of a *Broad Holloway*.

FIGURE 4.5 Dynamic stability along the 'Broad Holloway'.

FIGURE 4.6 The 'Broad Holloway'.

This photo helps me to summarize the preconditions and thus formulate the purpose of my plan.

> *It is the task of the newly to be created science of* ***omniconomics*** *to lead the Earth and humanity away from a narrow path next to an abyss on the edge of a volcano, and to take us to a* ***Broad Holloway*** *along which we can move safely into the future.*

Recommended further reading (for details, consult the online Bibliography)

Book: Clarke, Arthur C. (1962/2000): *Profiles of the Future: An Inquiry into the Limits of the Possible.*

5

IMPETUS WORDS

Concepts that need to be thrown away

5.1 The birth of science

Science was born on a Saturday.

It was in the year 1687 when the *Philosophiæ Naturalis Principia Mathematica* was published. On the 5th of July (which was a Saturday). The most important work of Sir Isaac Newton – who wasn't yet a Sir at the time – can be loosely translated into English as the *Philosophy of Nature based on Mathematical Principles*. In this book, Newton formulated laws of nature for movement and forces, including gravity: laws of nature which are today taught in every high school in the world. It was a breakthrough: never before had a field of science been mapped so clearly based on *facts*. Observations.

These observations had been performed systematically, not just by Newton himself, but also, primarily, by his predecessors, such as Tycho Brahe, Johannes Kepler, and Galileo Galilei. They studied the movements of the planets (including the Earth) and moons (including ours). Newton, who is by many considered the greatest genius of all time, used all results of these observations to formulate his famous laws of movement: we still refer to them as 'Newton's laws of motion'. He also introduced the concept of 'gravity', which enabled him to calculate the movements of planets and moons and at the same time, also the movements of falling or thrown objects on Earth. He could even predict these movements, using the mathematical formulas he had created.

Thus the oldest science was born: the science of physics and astronomy.

Why were physics and astronomy the first science? Why not biology or economics, for example? For a very simple reason: the field of physics is the *easiest*. Whoever reads a modern book on quantum mechanics will perhaps not be convinced, but the early areas of physics concern themselves with very simple things. It's about a planet moving around the Sun once a year, where both the planet and the sun are,

DOI: 10.4324/9781003306269-6

for simplicity, considered as dots without structure, moving around in a free space. Or about a ball leaving a cannon, moving towards city walls in need of crushing. The trajectory of the planet is ellipse-shaped. That of the cannonball is a parabola,* and we've known this for hundreds of years now, thanks to Newton. These processes are very simple, when compared to the way a plant grows. A planet, described as a dot, is much simpler than a flower bud or a human being.

Newton's book is often simply called the *Principia*. But the full title shows that Newton himself did not speak of science but of *natural philosophy*, which demonstrates the origins of this new form of human creativity: in philosophy. So, in one way or another, science has developed from philosophy. How this happened, is useful educational material for economists.

Naturally, there had been discussions about issues in the area of physics for thousands of years. By Aristotle (384–322 B.C.), for instance, who wrote in his book *Metaphysica* (around 330 B.C.E.) that the planets moved in circles because this was their 'natural behavior', as they existed in 'divine spheres'. This concept survived proudly for the entire Middle Ages. In those same Middle Ages, experts on cannons suspected that the trajectories of their cannonballs consisted of three parts: the first part tilted upwards in a straight line, the next described the arc of a circle, and finally there was a drop straight down. Thanks to this misconception many city walls will have been spared.

These claims were not systematically tested. This should not come as a surprise, as they lacked not only the right equipment, but also the interest. Testing hypotheses was primarily done by the head and not with the eyes: intuition and a sense of logic, rather than observations, dictated the evaluation.

For that reason, the physics of the classical era and the Middle Ages are not to be considered as science but as philosophy: *metaphysics*.

During and after the Renaissance, this changed gradually. Almost everyone knows the story of Copernicus (1473–1543), who wrote that the planets do move in circles, but around the sun instead of the earth.† And of Galilei (1564–1642), who claimed the same, and ended up receiving ecclesiastical punishment. It was also Galilei who first aimed a telescope at the sky and discovered the craters on the Moon, the moons of Jupiter and the rings around Saturn. 'Ansae', he called them: 'handles'.

Around the same time, the quality of the instruments increased greatly. This resulted in sophisticated astronomical devices: besides the telescope, also the quadrant and the sextant, the parallacticum and a greatly improved astrolabe were developed (Figure 5.1). And, of great importance: accurate clockworks. Using these, much more reliable observations could be made, as proven by the Dane Tycho Brahe (1546–1601).

* That is, if friction by air may be ignored, which on our planet does not always work well.

† For reasons of simplicity, the complicated explanation about the epicycles – circles upon circles upon circles – is left out.

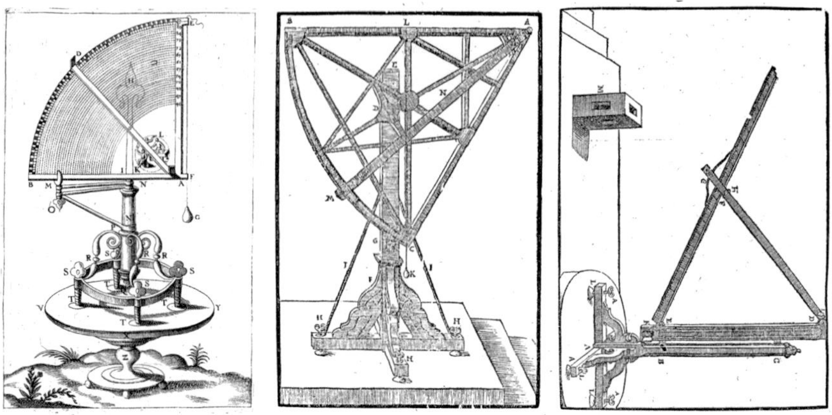

FIGURE 5.1 Quadrant, sextant and parallacticum, drawn by Tycho Brahe (published in a posthumous edition)[21].

Math also greatly improved around this time. It was Johannes Kepler (1571–1630), an assistant of Brahe, who applied the new mathematical methods on the latest observational data, after he had stolen them from his employer. One of his sensational conclusions was that planets don't move in circles at all, as their trajectory has the shape of an ellipse.

5.2 17th century, physics and astronomy: impetus

Do you know the word *'impetus'*? There is every chance you don't – at least, not in the context of physics. Which is no wonder, as it is no longer used there in our time. The word has been thrown away.

Since the word nevertheless plays an important role in this book, it is useful if you know how it is pronounced. The emphasis is on the first syllable, the 'e' sounds as in 'the' and the 'u' as in 'push'.

In late medieval physics, up to Newton, impetus was a core concept. People had puzzled for a long time, at least since Aristotle, about what was going on with objects that were moving. Were they perhaps 'animated', 'charged', or perhaps 'infected' with something? If so, with what?

In the 6th century AD, Joannes Philoponus gave as an answer: with *immaterial movement enérgeia.*[22] He owed the word *enérgeia* to Aristotle; it means something like activeness. Examples of enérgeia are pleasure, happiness and movement.

In our current time, we are inclined to say that the issue of the nature of moving objects has not been solved by putting a name on it, but Philoponus was probably very satisfied with it. Which was no wonder, because such a thing happened before and during the Middle Ages, and even now sometimes. Word invented; label stuck; problem solved.

Philoponus was quoted in the 11th century by Avicenna, who explained that an object that is set into motion has a 'tendency'.[23] In the 12th century, Nur ad-Din al-Bitruji added a 'desire'[24] to explain how the movement from a First Mover in the Heavenly 9th Sphere was transferred to the other spheres, so that all the activities that we now observe were set in motion.

In the 14th century, Jean Buridan brought the discussion from the Arab world to Christian Western Europe. He conceived the word *impetus* as a 'movement-maintaining' property of objects after they have been set in motion by an external cause. Impetus is, he stated in or around the year 1362, a force that allows an object to move in the direction in which the mover has steered it: up, down, sideways or in a circle. Giambattista Benedetti made a correction in 1585, as he excluded circles and restricted impetus to linear motions.

This was roughly the state of the discussion that Newton encountered when he began to design his physical theories in the 17th century. It was all still philosophy: about the (lifeless) nature, hence the justified name 'natural philosophy'.

Yet, the discussion progressed notably. According to Aristotle, who was regarded as an unquestionable authority for more than a thousand years with regard to natural philosophy, it is the natural tendency of objects in the *sublunary* realm (i.e., on Earth) to slow down and come to a standstill; objects that remain in motion require a constant cause of motion. On the other hand, in the heavens it is the natural behavior of objects to keep moving in circles without slowing down. But gradually, through the discussions of the above gentlemen, the realization came through that both statements were incorrect. It is the natural tendency of objects both on Earth and (according to Benedetti) in the cosmos to move in a straight line and to continue to do so, because these objects possess *impetus*. It is precisely when objects are slowing down, deflecting or even moving in a circle that an external cause is needed.

As you can see, one natural philosopher explained 'impetus' as an *enérgeia*, which is a bit like our word 'energy' but does not have entirely the same meaning. Another spoke of a power, and a third called it a 'tendency'. The scholars were clearly still searching, in a quest that yet had to lead to a real understanding of what motion was exactly. In the last centuries before Newton brought clarity, they came close: they were getting 'warm', as we would say.

It was Newton who was more than just 'warm' when he discovered that 'impetus' is actually a mix of a number of different concepts, including: force; energy; momentum; and inertia. When Newton split up this impetus into these different concepts, he succeeded in designing mathematical formulas that really worked, that is to say: with which he could not only understand motions but also predict them. Because Newton knew the right words, he was able to say the right things.That is why he remarked soberly: '*Hypotheses non fingo*'[25]: I do not make up hypotheses. With which he wished to say: my formulas are derived from and tested against reality, and it appears to be correct. That is the reason why Newton's work is regarded as the beginning of science. (See Appendix 2 for a further explanation.) After that, the word 'impetus' became superfluous and was discarded.

In the last centuries before Newton, thinking was in a transitional phase. It was no longer pure philosophy like that of people like Aristotle and Avicenna, what they were doing. But it was not yet a genuine science. People were 'warm', as someone who is looking for the soon-rising Sun in the morning twilight. It was the transition from philosophy to science. It was *protoscience*.

Protoscience is characterized by a number of well-recognizable characteristics. I will show you a list of them soon. One of them is: the use of words with which – in retrospect – you are 'warm', you come close, but you are not yet quite right. You will see that every science, in its proto-phase, possesses such words. I call them *impetus words*. Here is a short description. See Appendix 3 for a more formal definition.

An **impetus word** is

A word that does not yet address the essence of what a particular discipline entails. It is a diffuse concept, the use of which can lead to confusion or even to damage, because of a vague or even paradoxical meaning, which in retrospect may be in the neighborhood of a more successful term or even a mixture of several such terms. With advancing insight, the word will be discarded, or at least redefined.

5.3 18th century and beyond: chemistry, biology, medicine, neuropsychology

Chemistry, 18th century: Phlogiston

Just like *physics* arose in the 17th century from *metaphysics*, a branch of philosophy, in the 18th century the next science, *chemistry*, arose from *alchemy*. This was – looking back – a nice mix of philosophy, religion and magical thinking. A fascinating impetus-like issue took place in protochemistry around the question of what exactly happens when things *burn*.

In 1669, Johann Joachim Becher recorded his suspicion that substances that are flammable consist of a mixture of ash and *terra pinguis* (literally: fatty earth). During the combustion process, the latter escapes into the air and the ash remains.

At the beginning of the 18th century, Georg Ernst Stahl adopted this hypothesis and expanded it: according to him, the rusting of metal was essentially the same process as burning, only much slower. He had seen it well: he was 'warm'. He called the escaping substance *phlogiston*.[26]

Then came Antoine Lavoisier, who is widely regarded as the father of modern chemistry. In the years between 1770 and 1790 he showed that Stahl's assumption cannot be correct, because during the rusting, metal does not become lighter; indeed it does not even remain equal in weight: it becomes heavier. Lavoisier showed on the basis of a long series of experiments that combustion is not about a substance that *escapes* but rather one that is *caught*, namely: oxygen.[27] Phlogiston was a nice hypothesis, which was only exactly 180° wrong: a kind of 'anti-oxygen'. Thanks to Lavoisier, phlogiston no longer occurs in chemistry: it is an impetus word, it is abolished.

Biology, 19th century: Odyle

I bet you also don't know the word '*odyle*', not even in one of the other forms used, such as '*odic force*' or '*önd*'. It is an impetus word from 19th-century protobiology; it has now disappeared. And yet quite a few researchers of nature knew without any doubt: living, and even inanimate yet organic matter was no ordinary matter; there was something extra. This was the widely accepted position before biology became a mature science in the 19th century. It was Carl Reichenbach (1788–1869) who introduced the word 'odyle'[28]: he derived the word from the name of the supreme god Odin. His theory, which is called 'odology', was later appropriated by theosophy and thus ended up (or stayed) in magical-religious thinking.[29] Reichenbach was not the only one who thought of some kind of vital essence with which organic substances are penetrated. Caspar Friedrich Wolff (1733–1794) spoke of '*vis essentialis*' (essential force). Johann Friedrich Blumenbach (1752–1840) used the term '*Bildungstrieb*' (shaping urge). Their various theories are collectively called *vitalism*.

Jöns Jakob Berzelius was also a vitalist. At the beginning of the 19th century, he concluded from his doctrines that it should be impossible to produce organic substances from purely inorganic ones. But that went wrong, because a few years later urea, a component of urine, was indeed composed of inorganic substances. When more and more organic materials were manufactured in the chemical lab, Berzelius' view could not be maintained. Of course, not everyone spontaneously gave up vitalism; usually, obsolete theories only die out when their most fanatical supporters first do that themselves – and that took a while. Even in 1907 Henri Bergson did another vitalist attempt. His '*élan vital*' has proved to be a keeper: for $60 €52 you can buy a pot of 180 pills.

Medicine, 20th century: Hysteria

For the sake of completeness, I do not want to skip medicine. For physics and astronomy became a science in the 17th century, chemistry in the 18th and biology in the 19th century. A nice list, to which you can add medicine in the 20th century. A fantastic impetus word from its proto phase is '*hysteria*', a word that has been around for a few thousand years, but was seen by the end of the 19th century as an ailment in women which had unrest as its main symptom. The disease was controlled with the aid of genital massage, first manually by doctors and midwives, later with an electrical device. As the intriguing movie *Hysteria* by director Nadya Wexler shows, this electric development later turned out to be the invention of the vibrator.[30] Since medical science recognizes the female orgasm, hysteria has been abolished in the above sense.

Geology: Geosynclinal

Geology also became a mature science in the 20th century, after an important fundament was laid, based of the work of Alfred Wegener in 1915 and his predecessors, in the form of the phenomenon of the continental drift explained by plate tectonics, describing how the great plates slide over the Earth's surface, taking with them the continents. Before the causes became clear, there was a nice impetus word, 'geosynclinal', together with its counterpart 'geanticlinal', based on theories about a shrinking planet Earth, causing wrinkles (like on a drying apple) to lead to mountain formation.[31] I will not explain this further; if you want to know more, there are plenty of sources available on the internet.

Neuropsychology, 21st century: free will, me

And then a new mature science is approaching! At least, that's what I suspect. It is currently being born from the merger of two areas: psychology and neurology. I expect *neuropsychology* to give us a better understanding of what it is to be a human being. This will include putting an end to the impetus word '*free will*'. This term produces a remarkable paradox: on the one hand, neurologists manage to 'prove' that people do not have free will: our behavior results from automatic neurological processes; our decisions have already been made before our will comes into play.[32] On the other hand, a huge number of people, including psychologists and lawyers, insist that we do have free self-determination; if not, how can you ever declare someone accountable?

Actually, of course, this is also a centuries-old philosophical and theological issue about the question of *predestination*: if everything is fixed in advance, then we never have anything to choose. If so, is it justifiable that some are condemned to Hell while others go to Heaven? (But if *not* everything is predetermined, how can God be omnipotent and omniscient?)

My suspicion is that neuropsychology in the course of the 21st century will solve the issue by abolishing the term 'free will' and instead designing new, better words that tell a narrative without such paradoxes.

The same then probably also applies to words such as *consciousness*, *unconsciousness* and *subconsciousness* and for Sigmund Freud's threesome *id*, *ego* and *superego*[33]: new sets of words will replace them too. Do not ask me which: they still have to be invented. Or are they already?

Even the words '*I*' and '*me*' will disappear from scientific language, since they don't do justice to the complexity of neurological and mental processes in our minds. Although they will certainly continue to be used in daily speech,* scientifically they are impetus words. I will return to this fascinating issue in Chapter 13.

* Although? You never know. All kinds of words have been labeled as insulting and replaced by euphemisms. If in the past you might be homeless, nowadays you are an 'urban outdoorsman'. Prostitutes turned into 'comfort women'. Mad-houses are called 'psychiatric institutions', the unemployed are 'between jobs' and the disabled are 'challenged'. We do not die anymore but 'pass away', which is the result of a 'negative patient outcome'. Once it is discovered that 'I' is offensive to people with a multiple personality disorder (formerly: split personality or, according to older insights, possessed), the word will be banished and replaced by a euphemism. Maybe 'we'?

Recognizing impetus words

Impetus, *phlogiston*, and *odyle*: these are words that science abolished long ago. It is easy to recognize them – in retrospect – as impetus words. The same is still true of *hysteria* and *geosynclinal*, although their removal from science took place a shorter time ago, after their impetus character was rock-solid. But *free will* and *I* have not been abolished at all; they still function abundantly in scientific models. In cases like this, where retrospective wisdom cannot be used, the proof has to come from other characteristics listed in the definition: impetus words cause confusion, paradoxes, contradictions and considerable harm. Demonstrating thus that a word is 'impetus', requires careful proofing. In the remainder of this chapter, I make a start on this; further proofs follow in the coming chapters.

5.4 Economics: value, profit, externality, ownership etc.

A number of impetus words have already been reviewed. Chapter 3, for example, dealt extensively with 'value' (page 22). Also 'profit' was already mentioned (page 24), just like words such as 'externality' (page 24) and 'market failure' (page 49). (See Appendix 3 for a further explanation.)

A separate case is the word 'human'. I don't want to say that it is an impetus word as such. But the field of economics has turned it into one, because the word is often interpreted there as the flat notion of the '*homo economicus*', or, in other contexts, as a synonym of 'human capital'.

There are undoubtedly many more words that are used as impetus words in economics. It is not my ambition to create a complete list of them. It seems to me much better if such a list emerges in the context of new, interdisciplinary omnicomomic scientific development.

By way of illustration, it seems good to put one more word under the magnifying glass, as a conclusion of the investigation into the protoscientific status of economics.

Property

With what right can you call something your 'property'? A lot has been written about that question in laws and treaties. In Appendix 3, I deal with some historical and ethical issues, which, however, have not really led to clarity, as an example shows.

The Republic of Groningen

The Dutch Civil War lasted only briefly: from 1967 to 1970. But the battle was bloody and the consequences were dramatic. In a period of just two and a half years, five to ten thousand soldiers were killed. Estimates of the number of civilian casualties even range from fifty thousand to two hundred thousand people: most of them due to hunger. In addition, half a million Groningers were driven away from hearth and home.

It all started with the discovery of natural gas in the province of Groningen. The Dutch government thought it was great, it saw mountains of gold. But the people in the northern Dutch province of Groningen thought: why, Dutch government? This is our natural gas! Our wealth.

Despite the objections, the production of the gas was taken in hand, thanks to a concession from the government in The Hague to Shell and Exxon, working together in a consortium, the Dutch Petroleum Company (NAM).

The dissatisfaction among the Groningers about the 'foreign' government of the Netherlands in general was a fact that had existed for some time. Groningers are, they say, 'ons oard minzen' ('our kind of people') with a distinctive character and an accent different than the rest of the Netherlands. It was therefore not entirely unexpected that the Groninger people claimed the right to collect the huge profits from the natural gas themselves. "Groningen is not a colony!" exclaimed the charismatic leader Fré Meis. "It's in our soil, so the yield is ours!"

Was it this claim that led the Dutch government to make proposals on provincial reclassification? The natural gas area would fall outside the Groningen administration, according to the plans that were announced on May 27 1967. That turned out to be too much for the furious Groningers. Three days later, the uprising broke out and the independent Republic of Groningen was proclaimed.

The Netherlands reacted forcefully. On July 6, 1967, 'police actions' began to control the rebellious province with brute force. This was a good thing for the Groningen government: the international press and various foreign powers saw the Netherlands as the aggressor. That provided a lot of support for the Groningers! France and Israel secretly began to provide the insurgents with weapons. There was diplomatic support from Spain, Portugal, Norway, South Africa and the Vatican. But Great Britain and the Soviet Union chose the other side and supported the Dutch government, also in the form of arms deliveries.

On August 9, the hastily established Groningen army went on the counterattack. The Netherlands reacted to the bloody violence in Groningen with a blockade of the rebellious region on land and in the sea: the brand-new republic was completely shut off from the outside world. The famine that followed in the months and years weighed heavily on the population; the people died like rats. Photos of extremely skinny children reached the world press. Food flights were carried out by the American Community to Keep Grunn Alive. Doctors without Borders sent nursing units for the countless injured. The battle was fiercely waged on both sides, and pogroms among the civilians were not unknown.

The narrative in this case really happened. Literally. Well, *almost* literally, because you have to adjust a few things. Where *natural gas* is mentioned, you should read *oil.*

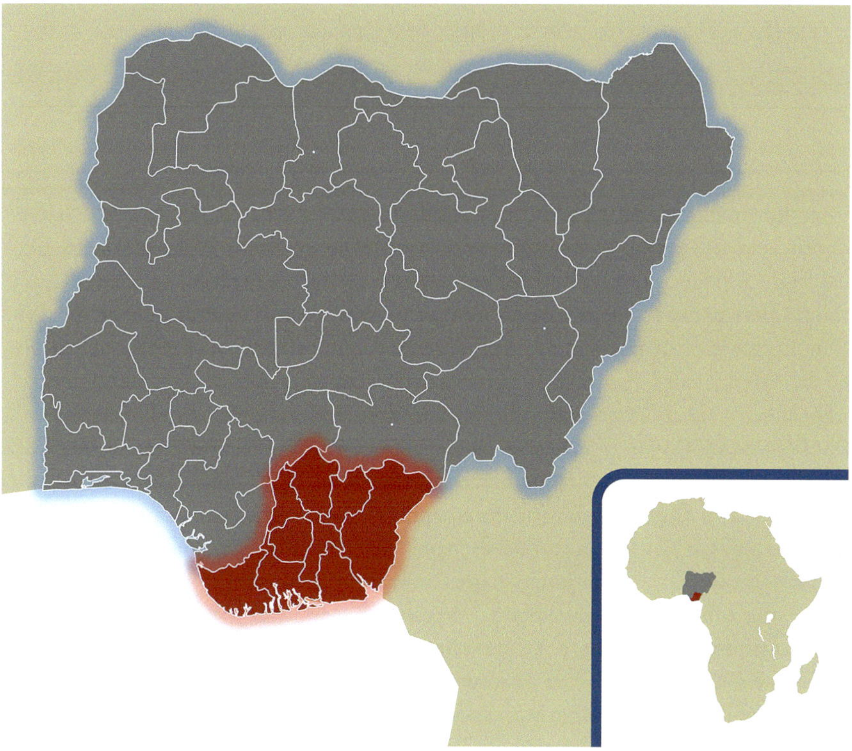

FIGURE 5.2 Nigeria and (in red) the rebellious region Biafra, in which the Igbo are the main population group.

Replace the *Netherlands* with *Nigeria*, the *Republic of Groningen* with the *Republic of Biafra* and *Groningers* with *Igbo*: see Figure 5.2 to orient yourself. The narrative has been moved to Europe for a while, because the western reader will experience it more easily as the unimaginably big drama it was.

Therefore, replace the Groningen Communist leader *Fré Meis* (who really advocated the Groningen natural gas rights) with the Biafran Colonel *Chukwuemeka Odumegwu Ojukwu*. If you then also replace Exxon with BP (but you can keep Shell) and *Dutch Petroleum Company* with *Petroleum Development Company*, then you are there.

Oh, and do not forget to multiply the number of victims (deaths, displaced persons) by ten.

How it ended?

> *Two years after the start, on July 19 1969, the 14th division of the Groningen Army launched a major offensive. Support was provided by Swedish mercenaries who together formed the Groningen Air Force of five pilots. They conquered an important city and advanced to the largest port city where they were stopped.*

> *On the morning of December 23 1969 Groningen was split in two by a large-scale Dutch attack. That appeared too much for the insurgents. After the final offensive on January 7 1970 by the 3rd Marines Division it was over. The surrender took place on January 14. The Netherlands was once again one country.*

In international law it has become self-evident that the mineral resources under a country are the property of that country. At first glance that might be perfectly logical, right and fair. But if you look better, there are still some problems. For example, the borders between the countries have often arisen fairly randomly. Now take the Dutch province of Groningen. From the very foundation of the Dutch *Republic of the Seven United Netherlands* around 1600, Groningen was one of the founding states of the Netherlands. But suppose that the struggle for independence of the Netherlands, which tried to separate itself from the powerful Spanish-German empire in the 16th and 17th centuries during the Eighty Years' War, was just as unsuccessful as that of Biafra (or the imaginary Groningen War of Independence): in that case, the natural gas proceeds would not go to The Hague but to Madrid; and that would have been just as logical and right according to the outside world. If, on the other hand, Groningen – because history was just a little different* – had not joined the Netherlands but the eastern adjacent diocese of Münster, the Berlin treasury would now be filled with an extra of more than €100 billion, at the expense of The Hague. But: both Madrid and Berlin, like The Hague, are not in Groningen. Incidentally, the city of Groningen, capital of the province of the same name, is not located in the natural gas area too, so …?

What about Biafra? Why should the Biafrans share 'their' petroleum with the much larger rest of Nigeria? The borders of the country stem from the colonial domination by European powers; they have hardly anything to do with the boundaries between peoples and cultures. Suppose that the entire Sahel (the sub-Saharan region, in which Nigeria is one of 15 countries) was left behind by the Europeans as one country: who's then would be the oil?

If the ownership and fair distribution of mineral resources between the existing countries is not easy, it is even more complicated if future generations are also involved. People who are not yet born don't have legal rights. But perhaps moral rights? In our time we take *all* easily recoverable fossil raw materials from the soil. We burn 95% of the oil directly and use the rest to produce abrasive materials and products. So, this can no longer be done by later generations, perhaps in new civilizations.† The metals that we dig up from open and other easily accessible mines are recyclable, but a real 100% recycling is impossible, since not one industrial or household process is perfect: no matter how carefully we will be, slowly but surely

* For instance, by a slightly different dissemination of the Reformation, so that around 1600 Groningen would have been predominantly Catholic instead of Protestant.

† Unless they are prepared to wait 350 million years to awake, so that there is enough time to create new fossil resources from life remains in the meantime.

the metals are getting so badly scattered and diluted over the world that they cannot be retrieved. Because of the current extraction, the quantity of easily accessible metals decreases, so that future generations will have to make more and more effort to get fresh ('virgin') material somewhere.

Proposals have been made to achieve a fairer distribution of scarce, once-only raw materials between current and future people. The International Resource Panel of the United Nations[34] calculated that the scarcest minerals are antimony and gold: according to an estimate by Henckens, their economic viability will come to an end around 2050.[35] Zinc and molybdenum are 'done' around 2100, followed by rhenium, copper and chromium (2150 to 2200). Tin, silver, lead, nickel and iron will be depleted in at most 300 years. It is possible that some of these metals will be used much more or less intensively, thanks to new technologies, and in that case the years will change drastically.

Henckens proposes that an international treaty be agreed, stating that for the scarce minerals a plan is designed to phase out the extraction step by step, with the aim of extending the extractability to at least a thousand years. According to his proposal, the scarce minerals are owned by all countries, both the countries that have them in their soil and the ones that import and use them. The user countries pay adequate compensation for this to the mineral countries. A system of fair distribution then allocates quotas to all countries, Henckens suggests.

Such a distribution agreement will be much fairer than the current distribution. Unfortunately, however, the experiences with existing distribution agreements are not very positive.

The Agreement Governing the Activities of States on the Moon and Other Celestial Bodies

(Better known as the Moon Treaty or Moon Agreement. Source: UN Office for Outer Space Affairs, 1979)

Article 11

1. *The moon and its natural resources are the common heritage of mankind, which finds its expression in the provisions of this Agreement and in particular in paragraph 5 of this article.*
2. *The moon is not subject to national appropriation by any claim of sovereignty, by means of use or occupation, or by any other means.*
3. *Neither the surface nor the subsurface of the moon, nor any part thereof or natural resources in place, shall become property of any State, international intergovernmental or non-governmental organization, national organization or non-governmental entity or of any natural person. The placement of personnel, space vehicles, equipment, facilities, stations and installations on or below the surface of the moon, including structures connected with its surface*

or subsurface, shall not create a right of ownership over the surface or the subsurface of the moon or any areas thereof.

4. *States Parties have the right to exploration and use of the moon without discrimination of any kind, on a basis of equality and in accordance with international law and the terms of this Agreement.*
5. *States Parties to this Agreement hereby undertake to establish an international regime, including appropriate procedures, to govern the exploitation of the natural resources of the moon as such exploitation is about to become feasible.*

Beautiful, isn't it? The Lunar Convention was negotiated from 1972 to 1979, in which year it was adopted by the General Assembly of the United Nations. After ratification by five countries (Philippines, Uruguay, Chile, the Netherlands and Austria) it came into force in 1984. However, because not one country ratified the treaty that practices manned space travel or has plans for it, the treaty has remained meaningless, although some still hope that it will once prove to be worthwhile.[36]

In short: a treaty like this does not work, because the countries that possess natural resources, or have the expectation to grab them earlier than others, simply keep the fountain pen in the breast pocket and the wallet closed.

The unfair distribution of natural resources between countries is a crucial cause of unsustainability, which is embedded in the deepest fundaments of the global political and economic system. The ownership of mineral resources, including the right to exploit them, will be extremely difficult to adjust. If we ever succeed in changing this, in a fair system the mineral resources will be the property of all humankind, equally spread over the entire human universe in three dimensions: north and south; east and west; present and future.

In order to achieve this, the current concept of 'ownership' will have to be thoroughly reviewed, legally, politically and economically. And there are perhaps more reasons for that. I now leave the property right to mineral resources and investigate the ownership of entirely different natural treasures.

Breast cancer

Meet: BRCA1, a gene in your chromosome #17. The detailed determination (*sequencing*) of its structure was completed in 1994 by the University of Utah, the US National Institute of Environmental Health Sciences (NIEHS) and the young commercial company Myriad Genetics. Even though it may be that you are not aware of it: you read and use this gene daily. Especially if you are a woman. Because the job of the gene is to protect your body against damaged cells, which could inadvertently cause breast cancer. The gene tells your body how the damaged cells can be repaired or – if that does not work – to kill them. As is often the case, the effect is determined by looking at what goes wrong when the gene itself is damaged.

You guessed it: the risk of breast cancer then is abnormally high. That immediately explains the name given to the gene, because BRCA1 is an abbreviation of *Breast Cancer* (1st variant).

Myriad Genetics and its partners applied for a patent immediately after their discovery, which they were granted four years later in the USA.[37] The patent not only granted exclusive rights to diagnostic, medical or pharmaceutical methods based on the knowledge of BRCA1, but *also the rights to the gene itself.* And that's weird.

It is weird, because patents can only be granted to individuals or companies that *first* possess the knowledge and applications that are to be patented, because they have developed them themselves (or have bought them from the developer). That is the essence of the entire patent laws! But Myriad and partners certainly did not have the knowledge about BRCA1 first. *You* have been in possession of the knowledge and skills of reading, writing and applying your DNA for as long as you live – even if you may do so unconsciously. You are not even the first, because our ancestors did the same millions of years ago, when we were not even humans. Each of us reads the genetic code every day, interprets the information successfully and applies it; if that were not the case, then humankind would have long ago succumbed to breast cancer and a thousand other ailments.* Companies that now claim and obtain the exclusive right of that information, commit theft.

It is therefore no wonder that in 2013 the Supreme Court of the United States concluded that this was crazy. In a famous judgment, the court ruled that BRCA1 cannot be patented.[38] "A naturally occurring DNA segment is a product of nature and is not patentable", the statement read. As a result, all patents already granted on genes from both humans and all other life forms, have been cancelled. This only concerns genes in their natural form, because genetically modified genes have been created by humans and are still patentable.

In Europe, patents on natural genes are still possible, and it is being done regularly.[39] So, if you are Italian, Polish or Dutch, you are being robbed of your heritage.

Both mineral treasures and genetic knowledge show: 'property' is a complex subject, with legal, political and economic significance. The current vision, enshrined in national and international legislation and treaties, contains a variety of problems. As long as they are not resolved, by replacing the word with either one or a number of other concepts, or by redefining it, it must be regarded as an impetus word. (See Appendix 3 for a further explanation.)

It is still too early to determine how 'property' can be reformulated, replaced or removed. For this purpose, it is first necessary to start with the development of omniconomic science, in the context of which appropriate new words can be designed. But perhaps the following text may be useful, as a source of inspiration to look at 'property' in a fundamentally different way.

* None of the nearly one thousand pages of European patent law state that knowledge and skills must be *consciously* owned or applied. You can check that yourself in three languages: EPO (2016).

The text is based on a speech that may have been given by the American native leader Si'ahl, head of the Duwamish tribe, in or around the year 1854. The chief is better known as Seathle or Seattle; the eponymous city has been named after him. The text of the speech was not written down until 1887 by Henry Smith,[40] who stated that the words had been spoken by the chief; however, it seems more likely that Smith composed the text himself.[41] Since then, the speech has been rewritten several times.[42] One of the versions comes from Ted Perry, who edited the text for a television movie called *Home*.[43]

Even though the text is far from authentic, the essence of the words of the original speaker or writer is probably still well represented. I quote some passages of the speech.

Man belongs to the Earth

"The great white chief Washington sends us word that he wants to buy our lands. But how can you buy or sell the sky, the warmth of the land? The idea is strange to us. If we do not own the freshness of the air and the sparkle of the water, how can you buy them?

Every part of this earth is sacred to my people. Every shining pine needle, every sandy shore, every mist in the dark woods, every clearing and humming insect is holy in the memory and experience of my people. The sap which courses through the trees carries the memories of the red man.

We are part of the earth and it is part of us. The perfumed flowers are our sisters; the deer, the horse, the great eagle, these are our brothers. The rocky crests, the juices in the meadows, the body heat of the pony, and man – all belong to the same family.

So, when the Great Chief in Washington sends word that he wishes to buy our land, he asks much of us.

The water's murmur is the voice of my father's father. The rivers are our brothers, they quench our thirst. The rivers carry our canoes, and feed our children. If we sell you our land, you must remember to teach your children that the rivers are our brothers, and yours, and you must henceforth give the rivers the kindness you would give any brother.

We know that the white man does not understand our ways. One portion of land is the same to him as the next, for he is a stranger who comes in the night and takes from the land whatever he needs.

The earth is not his brother, but his enemy, and when he has conquered it, he moves on. He leaves his fathers' graves behind and he does not care. His fathers' graves and his children's birthright are forgotten. He treats his mother, the earth, and his brother, the sky as things to be bought, plundered, sold like sheep or bright beads. His appetite will devour the earth and leave behind only desert.

If we decide to accept, I will make one condition: the white man must treat the beasts of this land as his brothers. What is man without the beasts? If the beasts were gone, men would die from a great loneliness of spirit. For whatever happens to the beasts, soon happens to man. All things are connected. Whatever befalls the earth, befalls the sons of the earth. If men spit upon the ground, they spit upon themselves.

This we know. The Earth does not belong to man. Man belongs to the Earth."

Recommended further reading (for details, consult the online Bibliography)

Book Addendum: Newton, Isaac (1713): *Scholium Generale.*
Motion picture: Wexler, Tanya (director, 2011): *Hysteria.*
PhD Thesis: Henckens, Theo (2016): *Managing raw materials scarcity: safeguarding the availability of geologically scarce mineral resources for future generations.*

6 PROTOSCIENCE

6.1 The seven characteristics of protoscience

I have used the word protoscience a number of times now. Protoscience is something like the twilight zone between philosophy and science. That is quite a vague description, and as this word will have a considerable part to play in this book, it is good to provide a further definition. Preferably, this definition is in alignment with the work of others.

The term protoscience is used by various authors, but often in a somewhat loose, vague way, as you can read in Appendix 3. Luckily one person provides us with some clarity. This is the famed science philosopher Thomas Kuhn, who has said and written a lot about the development of science. He is most famous for his description of the 'scientific revolutions',[44] of which the Copernican revolution, where Copernicus moved the center of the universe from the Earth to the Sun, is most commonly known.

Kuhn used the term 'protoscience' on only one occasion. At a congress held in 1970 in his honor, he noted[45]:

> "There are many fields – I shall call them proto-sciences – in which practice does generate testable conclusions but which nevertheless resemble philosophy and the arts rather than the established sciences in their developmental patterns. I think, for example, of fields like chemistry and electricity before the mid-eighteenth century, of the study of heredity and phylogeny before the mid-nineteenth, or of many of the social sciences today.
>
> In these fields, too, [...] incessant criticism and continual striving for a fresh start are primary forces, and need to be. No more than in philosophy and the arts, however, do they result in clear-cut progress. I conclude, in short, that the proto-sciences, like the arts and philosophy, lack some element

DOI: 10.4324/9781003306269-7

which, in the mature sciences, permits the more obvious forms of progress. It is not, however, anything that a methodological prescription can provide."

It is not easy to give a solid formal definition of 'protoscience' that is more concrete than Kuhn's. But even Kuhn's definition offers little guidance if you want to determine whether a discipline is a protoscience or a mature science.

Fortunately, there is another way to bring clarity, and that is by listing characteristics. With the help of such characteristics, which together make protoscience easily recognizable, an *operational* definition can be given; that is, a definition that can be used in practice to determine whether it is realistic to call a certain field of study 'proto-'. I give you my definition, for which the four spheres are useful.

A **protoscience** is

A field that is in a transitional phase between philosophy (or art, religion or magical thinking) and science, where one has not yet grasped the essence of what that discipline entails. A protoscience is recognizable because the field possesses all or most of the following seven characteristics:

Logosphere

1. ***Struggling Schools:*** There is a multitude of schools, which are in competition about the key points of the field. This shows that the essence of the field has not yet been grasped.
2. ***Impetus Words:*** The prevailing paradigms in those schools are – at least partially – impetus words.

Mythosphere

3. ***Myths:*** Using these impetus words, questions are asked and narratives are told ('myths'), which are confusing, vague, contradictory, paradoxical or simply proven to be false.

Politosphere

4. ***Copied Methods:*** Methods and techniques are used that are successful in other fields, but less so in this field.
5. ***Oversimplification:*** While doing so, reality is oversimplified.

Phenosphere

6. ***Lack of Empirics:*** Experiments and other methods of empirical testing are not performed or are performed infrequently; and to the extent that they

are, the results are not or poorly reproducible. There is little need for replicability.

7. **Lack of Success:** There is no or little predictive power, or any other kind of 'success'.

I've already written about impetus words. I will now take a closer look at the other characteristics.

Protoscience is not equal to pseudoscience

A small side note. It is important to note that protoscience should not be confused with *pseudoscience*. Protoscience is potentially a precursor of real science, while pseudoscience is un- or even anti-scientific nonsense that is deliberately disguised as science:[46] superstition, quackery and in many cases a conscious attempt at manipulation and deception. Pseudoscience is practiced on the one hand on the basis of uncritical forms of belief or superstition, such as creationism and homeopathy (see Chapter 14). And on the other hand, by or on behalf of parties that try to defend commercial interests.

6.2 Schools

In 1970, Kuhn named the study of heredity and evolution before the mid-19th century as an example of protoscience. This is a good starting point for examining property number 1: the existence of schools. I will start with a former protoscience that has since become fully mature.

Schools in biology

In 19th-century biology, besides the issue of the mysterious life force, there was another important issue. As more and more fossils were found from plants and animals that were no longer seen alive on the surface of the Earth, the question became increasingly urgent, where the natural species diversity comes from. Of course, there were various 'schools'. This is an important subject in preparation for the narrative about schools in economics, so I give it ample attention.

For example, Joseph Gottlieb Kölreuter launched the idea that species change thanks to a process called '*transmutation*' as a result of *hybridization*, i.e. by mixing species.[47] The transmutationist school became quite popular among protobiologists.

Another famous nature researcher, Jean Baptiste Lamarck, used a slightly different term: '*transformation*'.[48] According to him and his followers in the 'Lamarckist' or 'transformationist school', simple life forms from inanimate matter arise through '*generatio spontanea*', spontaneous generation, after which a vitalist life force pushes the species 'upwards' in the direction of increasing complexity. Properties obtained by individual specimens of a species as a result of adaptation to the environment can be directly passed on to the offspring – according to the Lamarckists. Building on the transformationist school, Robert Knox and Robert Grant developed their own 'Edinburgh school', which emphasized similarities in body shape or structure between species ('homology') as an indication that they had common ancestors. Their view was a bit too simple, but 'warm'.

Charles Babbage advanced another hypothesis.[49] During God's creation, with the help of His omnipotence and omniscience, God created natural laws that allow new species to emerge at appropriate times in the course of world history.

To Robert Chambers, we owe the *orthodox* or *teleological* school, which suspects that species arise according to an 'orthogenetic' (aiming right from the beginning toward the final purpose) plan that has as its ultimate goal the origin of man.[50]

Like Chambers, geologist Charles Lyell surmised that species are unchangeable, but he believed in a kind of continually progressing creation process of disappearing and newly appearing species, not on the basis of a divine plan but because living environments are changing.[51] Another early scientist, Georges Cuvier, also insisted that species are unchangeable; according to him, that is because their parts are too finely tuned to each other; thus, every small change would prove disastrous.[52] We still find this misunderstanding with creationists, who continue to try to argue in this way that all living species were designed and created by a god or a superintelligence,[53] despite careful rebuttals of such an unchangeable finetuning.[54]

And there was yet another group that opposed transmutation. These were the *naturalists*, influenced by German supporters of the philosophical movement called *idealism*, to which Goethe and Hegel belonged. Naturalists Louis Agassiz and Richard Owen were convinced that species could not change because each species is the embodiment of an 'idea' in the mind of the Creator.

All these schools cheerfully mixed scientific, philosophical and religious thoughts and drew conclusions that could not be verified empirically. Until the evidence accumulated in the course of the 19th century that species really do change; in fact: that the history of our planet is a coming and going of evolving species. Anyway, you know how that ended. Charles Darwin published his famous book *On the Origin of Species* in 1859, followed by a few other important works that together brought biology into the mature scientific phase. In this science, evolution has become the major connecting theme of the entire discipline. The impetus words *odyle*, *transmutation* and *orthogenesis* have disappeared through the side door.

Schools in economics

It would be a nice exercise to count the schools in protobiology that I have mentioned. I didn't do that, but I am sure that this number is nothing compared to the number of schools in protoeconomics. And I am not talking about the economy of the 18th or 19th century. I'm talking about *now*.

There are dozens of them. I made an effort to compose a graphical overview of them. You see the result in Figure 6.1. For this purpose, I used a range of sources.[55] But it was not easy, because the various sources do not agree very much. In the first place about what exactly is an economic school and what is not; and secondly, about which school originated from which earlier one, or perhaps from a combination or a synthesis of several schools. So, Figure 6.1 is my own subjective interpretation. Therefore, don't consider the figure as an absolute truth, but rather as a rough sketch of economic theoretical reality.

I can't help it: when I look at Figure 6.1, the comparison with the history of the religious denominations in and outside the Netherlands arises. You could make a similar scheme – which would be just as much subjective – in which Roman

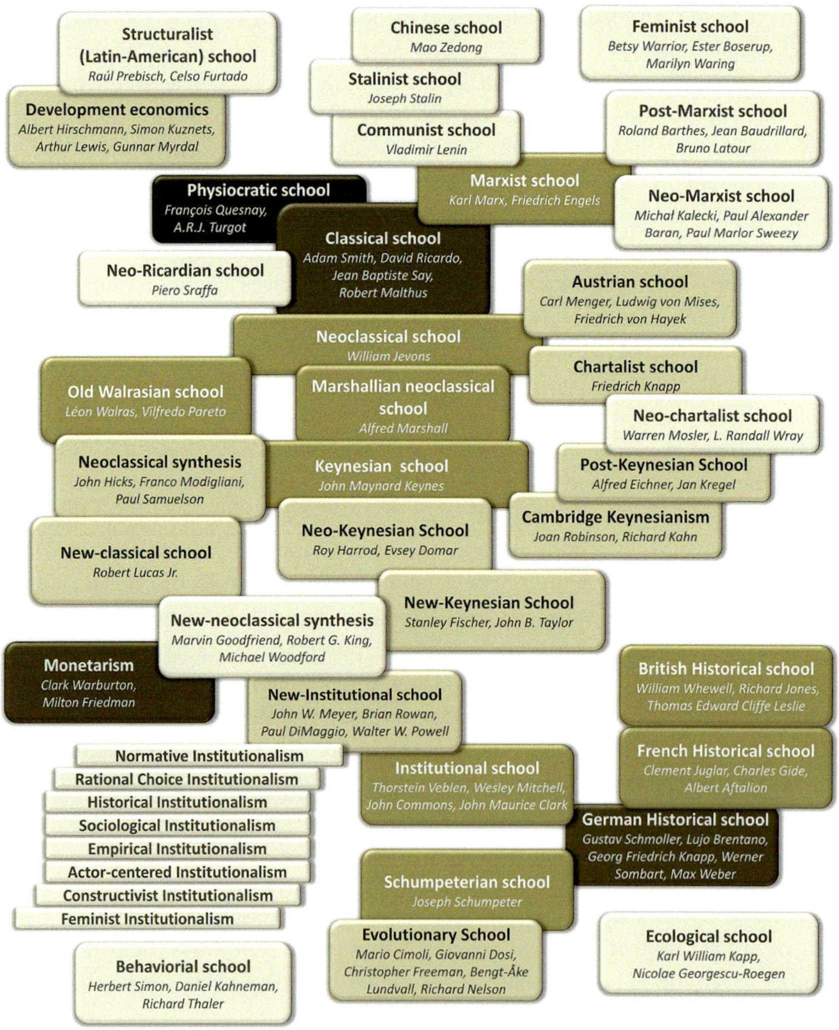

FIGURE 6.1 Some economic schools.

Catholic, Lutheran and Calvinist belong to the core schools, which then split into a mass of large and small churches and splinter groups. That scheme also has its syntheses, for example the Protestant Church in the Netherlands. You know what happens when two churches merge? A few days later you have *three*.

You see in the figure a collection of eight forms of the *New-Institutional school*: from *normative* to *feminist*. You may see that list as an illustration; I could have placed a similar series at other schools.

Every religious denomination and every economic school has their own narrative. With help from a number of experts, I summarize some of them.[56]

Short narratives from some schools

According to the **Neoclassical** narrative, the economy consists of individuals who are completely egoistic and always act rationally. Economy is mainly about trade and consumption, and primarily changes as a result of individual decisions.

Recommendation: *free market with modest government influence.*

The narrative of the **Austrian** school looks similar, but places more emphasis on tradition as a source of human action.

Recommendation: *give the free market all space!*

However, the **Marxist** narrative tells that the economy is made up of classes. People are selfish and rational, except when workers struggle for socialism. Economy is mainly about production; changes come from class struggle, capital growth and technological progress.

Recommendation: *socialist revolution followed by central planning.*

In the **Keynesian** school the emphasis is on classes too. People are not very rational; they are driven by habits and instincts. The Keynesians disagree about the role of selfishness, about what is the main subject of the economy (production, according to some) and about economic development.

Recommendation: *active fiscal policy aimed at redistribution for the purpose of poverty reduction.*

According to the narrative of the **Institutional** school, the economy – not surprisingly – consists of institutions, and also of individuals. Human behavior comes from various mental layers: instinct, habit, beliefs, ratio. More than with the neoclassical school, emphasis is placed on production. Economic changes (naturally) arise from interactions between individuals and institutions.

The institutionalists disagree about recommendations *for policy.*

The **Behaviorists** have the richest economy, because for them it consists not only of individuals and institutions but also of organizations. The stratification of the human behavior, described by the institutionalists, is recognized by the behaviorists to a limited extent.

Recommendation: *not necessarily against government intervention.*

The **Schumpeterian**-oriented school likes to talk about non-rational entrepreneurship as a motive for human behavior. According to them, economy is about production; development is the result of technological innovation.

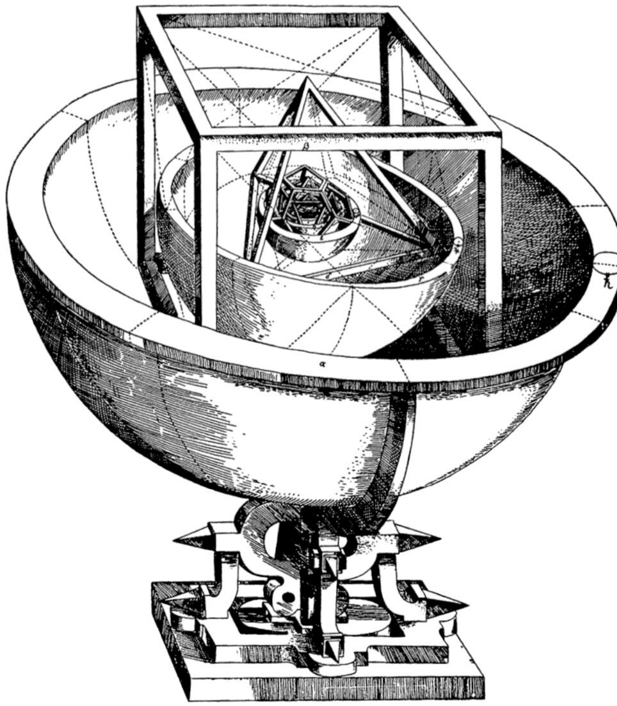

FIGURE 6.3 Kepler's Cosmic Cup.

Proudly, Kepler wrote about his discovery in 1596 in his book *Mysterium Cosmographicum*. In it he made a sketch of his model, which he called the 'Cosmic Cup'. You can see the sketch in Figure 6.3. Unfortunately, that is all that happened, because Kepler's efforts to raise funds for a real silver cup were unsuccessful, so I cannot show you a photo of it.

Ah, how unfortunate is it that the cosmic plan was cruelly disturbed, by Kepler's own hand! Because yes, later he proved, as said, that the orbits are not circular but elliptical. But hey, this was still the protoscientific time, wasn't it? Johannes found an easy way out. With some (metaphorical) cutting, sawing and gluing he saw an opportunity to reconstruct the Cup. In 1621, he wrote a new version of his book, but only a few could share his enthusiasm. That was no wonder: The Cosmic Cup was a fairy tale. A beautiful myth.

Pattern recognition in modern times: the face

Kepler had, he thought, discovered a special pattern in the universe. He was not the only one who enthusiastically practiced pattern recognition. Because we all love to do it: recognize patterns. Thus, all people have a special talent for discovering *faces*. Not only when we see people, even in inanimate objects. This skill is called '*pareidolia*'. A famous example is the 'face on Mars', which was photographed in 1976 by a NASA satellite. The photograph, shown in Figure 6.4, caused quite a pseudoscientific stir: some saw in the face evidence of a long-lost Martian civilization.[58] A new, much sharper photograph from 2007 put an end to the sensation by showing that it was an ordinary rock formation. It turned out to be an optical illusion, a myth, not originating in reality but in perception.

Pattern recognition in protoeconomics: cycles

Economists are equally eager to find special patterns. This is all the more true when it comes to the economic conjuncture, moving up and down, because a lot depends on that. Sometimes the economy goes up and everything is great, but a little later it all collapses, stock prices go down (gradually or dramatically fast) and all kinds of companies go bankrupt. Who would not want to understand these phenomena, or better: predict or even prevent them?

In order to show you which principle is used to discover patterns in at first sight chaotic conjunctural movements, I first introduce: *adding up* graphs. You can see an example of this in Figure 6.5. In the upper half, three fictional cycles are depicted. They are *harmonic*: they go up and down in a flowing, mathematically precise way (as *sinusoids*) over the 220 years of the graph. Wave #1 is the slowest: its period is 220 years and its amplitude, its maximum deviation from the average, is 9.

You add the three cycles by reading the height of each cycle for every separate year – shown on the horizontal axis – adding the three numbers and plotting the result in a new graph. The result is shown in the bottom half of Figure 6.5. In it, you can still clearly recognize the shape of Wave #1, and that also applies to Waves #2 and #3.

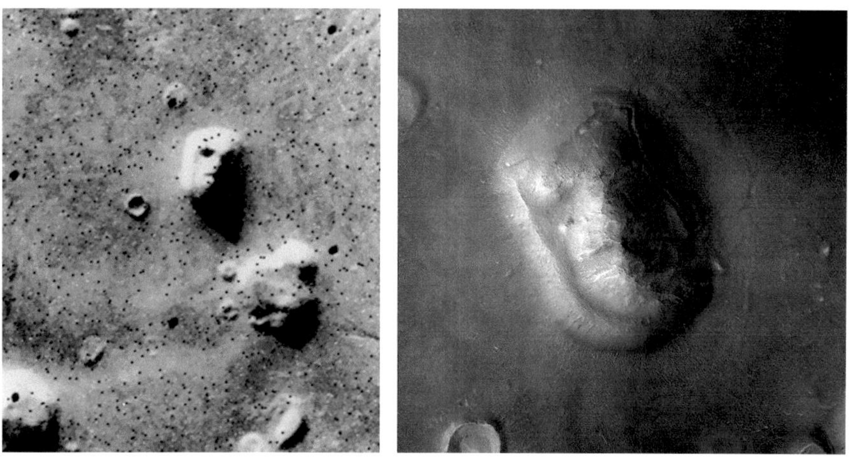

FIGURE 6.4 The face on Mars. Photo left: NASA, Viking 1 Orbiter, 1976. Right: NASA, Mars Reconnaissance Orbiter, 2007.

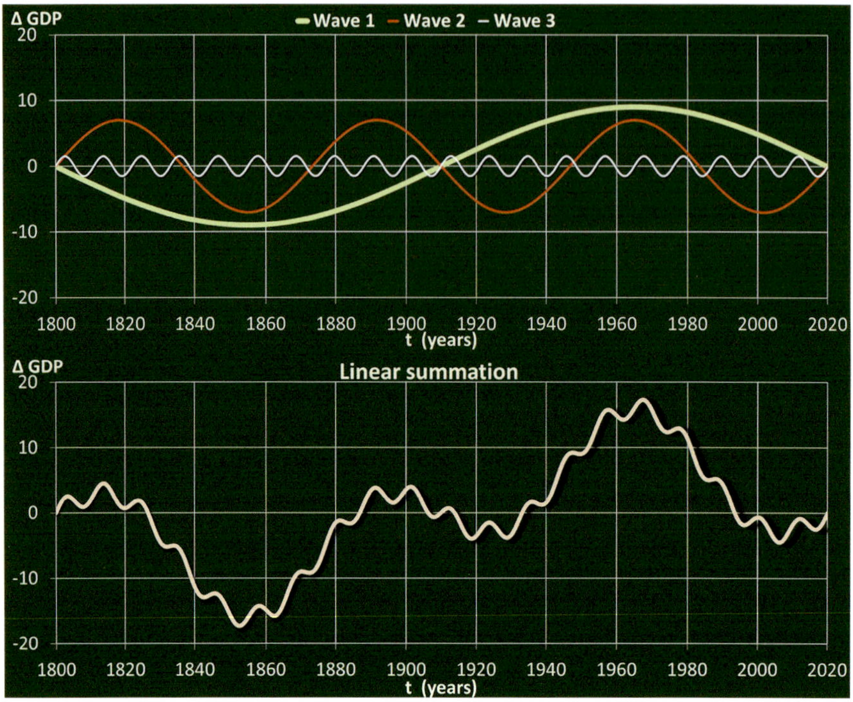

FIGURE 6.5 The addition of harmonic waves.

As said, the bottom graph is obtained by simply adding the values of the three cycles. *Linear* addition, as it is called. This is only allowed if you have sufficient reason to assume that the three cycles are mutually independent, that is to say that they do not disturb each other's pattern. This is the case with *sound waves*, which is why you can easily recognize the three separate voices in a combination of guitar, piano and a human voice. If this is not the case, you cannot add the cycles linearly. At best, more complex mathematical operations are involved that are highly dependent on the situation. In many practical cases, there is no valid way to merge mutually disruptive cycles.

Now, economists are not trying to merge cycles. They have the opposite task: to split up a whimsical development into various neat cycles. A typical example is shown in Figure 6.6. This graph shows the development of the gross domestic product (GDP) over the past two centuries. To be precise, you do not see the GDP itself, but its increases and decreases (ΔGDP) for the United States from 1800. The choice of this graph is not arbitrary, because economists consider the increase or decrease in GDP to be a crucial indicator of the state of the economy. Where the curve is above the zero line, the GDP rises; where it is below, it falls. Since the graph is on average more above than below the zero line, GDP grows over the years.

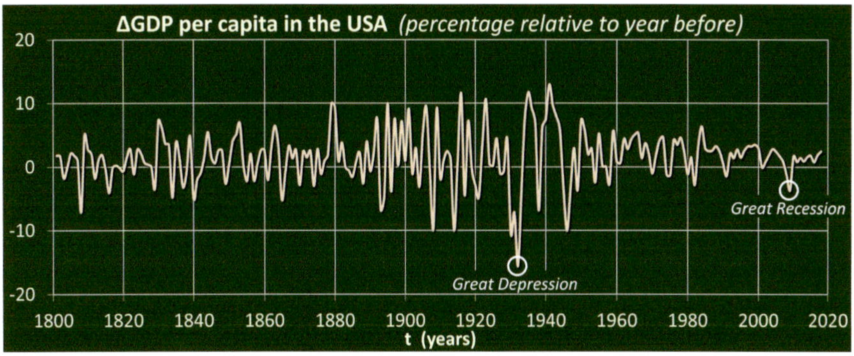

FIGURE 6.6 Increases and decreases in GDP per capita, 1800 to 2015.[59]

The figure is extremely capricious, and at first sight it seems impossible to recognize a pattern in it. But economists don't only have this one chart, they have thousands of them, since you can look at many other phenomena: stock prices, trade prices of all sorts of individual commodities and products, corporate profits, incomes, you name it.

Economists believe they have discovered a pattern for a number of such separate economic phenomena. For example, it is argued that a cycle can be recognized in the investments in *infrastructure*. This cycle is called the '*Kuznets swing*'. This swing also covers other economic developments, according to economists, including the well-known *hog cycle*.

They also discovered the '*Juglar cycle*', which concerns loans for industrial investments. And there is the '*Kondratieff wave*', which relates to *everything*, that is, to the economic conjuncture as a whole. In total, five cycles were observed, each with their own rhythm. They move simultaneously, right through each other.

It is very clever! To distinguish five such fluctuations, amidst thousands of different variables, many of which move up and down wildly. In order to demonstrate credibly five periodic developments interfering with each other, one would expect (a) that the five waves are fairly *independent* of each other, because if they influence each other mutually, it becomes a mess and they will never be separately discoverable. If you nevertheless recognize them, you see ghosts. And (b) that their periods and amplitudes are quite *stable*, otherwise you can really reason in all directions.

Unfortunately, these conditions have not been met in the slightest. To begin with, the postulated cycles can never be mutually independent. For example, the economic situation as a whole, described in the '*Kondratieff wave*', has, by definition, an impact on *all* economic phenomena, including on the other cycles. In addition, it is difficult to imagine that cycles in investments in, for example, industry and infrastructure can be completely independent of each other.

As for the periods of the five cycles: here is an overview.

1. *Kitchin cycle: period = 3 to 5 years.*
2. *Juglar cycle: period = 7 to 11 years.*
3. *Kuznets swing: period = 15 to 25 years.*
4. *Kondratieff wave: period = 45 to 60 years.*
5. *Grand Super Cycle: period = 70 or more years.**

As you see, all five periods have a very large margin. This also applies to their amplitudes: one 'wave' is much more intense than the previous or the next. That way you can comfortably go in all directions. See also Appendix 4, where the intensity and capriciousness of the 'observed' waves is well illustrated.

In addition, the case is further muddied by the occurrence of other, non-periodic phenomena of which we are certain that they really exist. Among others:

6. *Trends that continuously work through: for example, technological innovations, increased prosperity, political developments, the rise or fall of economic superpowers, climate change.*
7. *Noise, that is: accidental changes, such as the weather in a given year.*
8. *One-off major incidents, for example world wars and pandemics.*

All this makes it very unlikely that the state of the economy can really be described in this way – let alone *explained*. But, in order to clarify matters further, I conducted a small experiment.

The wave experiment

By way of laboratory, I have designed a spreadsheet that produces results in the form of graphs. If you want, you can replicate the experiment yourself, because you can download the spreadsheet. If you don't know much about spreadsheets, you can easily use it to produce new graphs at the click of a button. If you are familiar with spreadsheets, you can even modify the calculation model as you see fit.

I have done it as follows. In the spreadsheet, the five cycles are first calculated separately. Since they all seem to fluctuate randomly in period and amplitude between wide limits, I mimicked that in the spreadsheet using the '*random*' function.

* Fernand Braudel (1949) believed he recognized an even longer cycle with a period of 300 to 400 years, leading to lows in Europe in the 6th, 10th, 14th and 17th centuries. His hypothesis received little follow-up and is not taken into consideration here.

Besides, I assumed and calculated some characteristics for each of the three non-periodic phenomena. Then I added up the five cycles and the three other phenomena.

By doing the calculation repeatedly, I get a different graph each time, thanks to the built-in randomness, and thus a nice variation of outcomes. One such outcome is shown in Figure 6.7. As in Figure 6.5, the top graph shows the five individual waves. The second graph shows the three non-periodic phenomena. The bottom

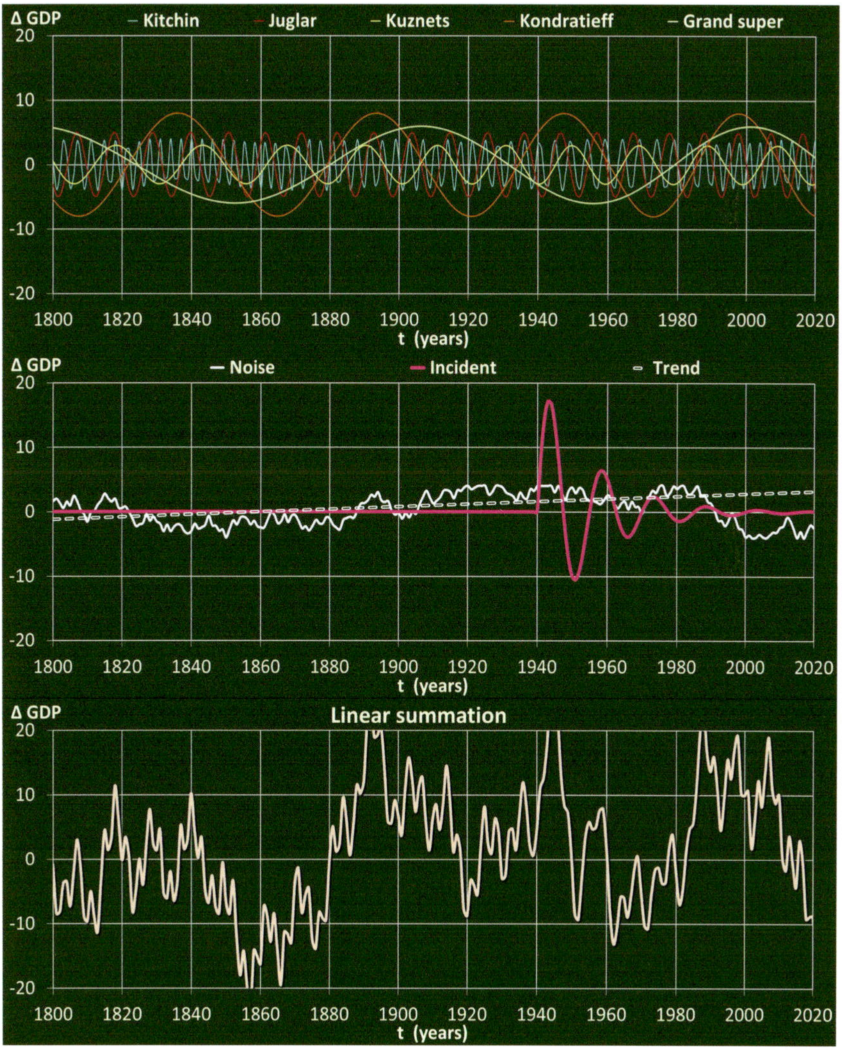

FIGURE 6.7 Linear addition (bottom graph) of the five cycles (top) and three non-periodic phenomena (middle).

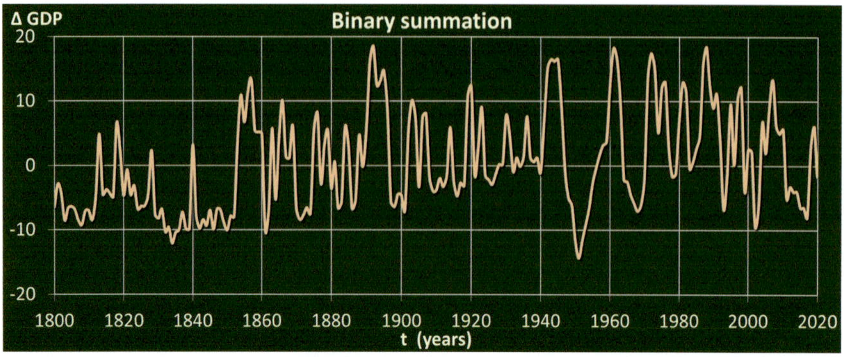

FIGURE 6.8 Binary addition of the same five cycles and three non-periodic phenomena.

graph shows the addition. For this I used the 'normal' linear addition, even though I know that this is not realistic.

For that reason, I also used a different method of addition, 'binary',[60] the result of which is shown in Figure 6.8. (You can find the details in the spreadsheet.)

Apart from these two ways of adding, there are countless other ways to combine the eight graphs mathematically. However, two addition methods are more than enough for me, because if you compare the two results, you can see that they differ enormously. Not only the randomness of the individual cycles, but also the freely selectable way of combining them offers unlimited possibilities.

My graphical experiment proves that you can produce an infinite number of different graphs, using the five cycles claimed by the economists. If you go on long enough, you can reproduce not only the observed graph of Figure 6.6 but really every other imaginable graph. The theory of the five cycles can explain *everything* and thus explains *nothing at all.*

The theorists

How does theoretical economics see this itself? For starters, it is interesting to look at a few proposed definitions of 'cycle'. In general, the basic definition states that a cycle is…[61]

> "… a type of fluctuation (…) [that] consists of expansions occurring at about the same time in many economic activities, followed by similarly general recessions, contractions and revivals which merge into the expansion phase of the next cycle. This sequence of changes is recurrent but not periodic; in duration business cycles vary from more than one year to ten or twelve years."

Not periodic! Indeed. But if you are talking about 'cycles' that are not periodical, what do you actually say? Just that every now and then something goes up and down again a while later. What do you have?

Okay, this definition is from 1946 and hence quite old. But a much more recent source does not say much else: "Business cycles are based on two parts: cyclical peaks and bottoms."[62]

Nevertheless econometrists, using a method widely used in physics called *Fourier analysis*, believe that the presence of real cycles can be demonstrated, which is remarkable, since this method only works with fixed periods and amplitudes.[63]

Discussions about the degree of reality of the economic waves are still ongoing. But where some think they recognize several of the mentioned waves in the available data, others do not see them and vice versa.[64] Rightly, Noah Smith wrote a publication entitled *"Seeing Economic Patterns That Don't Exist. That graph with regular peaks and valleys? Don't count on it being more than random noise"*.[65]

Economic theorists are – what I call – *eagercredulous*: they are so keen on seeing patterns that thorough, sober testing is largely absent. For good, critical scientists, this is not a good trait.

The 'Five Cycles' are a myth, no less than Kepler's Cosmic Cup and the Face on Mars.

This conclusion shows once again how easily you can fall into the trap of 'recognizing' patterns that have no meaning. The economic cycles fit beautifully in a row of nonsense patterns that have been discovered and told in an entertaining book by Tyler Vigen.[66] He proves, for instance, that there is a statistical correlation of no less than 95% between eating cheese and the number of fatal accidents in bed in the USA in the period from 2000 to 2009. In the same period, there is even a correlation of 99% between the consumption of margarine per person and the number of divorces (in the state of Maine). Better eat real butter?

After this joyous conclusion, it is time to explore some other myths around economic theories, starting with a remarkable category.

Recommended further reading (for details, consult the online Bibliography)

Book section: Kuhn, Thomas S. (1970): Reflections on my critics.
Book: Chang, Ha-Joon (2014): *Economics: A User's Guide.*
Book: Vigen, Tyler (2015): *Spurious Correlations. Correlation Does Not Equal Causation.*

7

ECONOMICS MYTHS

7.1 Automatisms

Automatically, the Sun rises. Automatically, heavy objects fall down from my hand when I release them. Those are *certainties*. Does economics also have such certainties?

Yes! many economists respond enthusiastically. I set out a number of their 'automatisms'.

> **The Invisible Hand Myth**
>
> *Fair competition in a free market automatically creates a natural order in society and an economic balance, as if an invisible hand always guards and restores that order.*

It was Adam Smith, considered the 'father of economics', who created this narrative (in 1776 and 1789). With that 'hand' he referred not to a deity but to an automatically occurring market mechanism. It became an adage of the 'Classical School'. By the way, Smith himself put this principle into a nuanced perspective; it was mainly his followers who absolutized the statement. Of course, this mainly concerned those who benefited greatly from entrepreneurship in a free market without any annoying intervention by an active government.

> **Say's Myth**
>
> *Every supply creates its own demand, and as a result the market is automatically always tailor-made.*

DOI: 10.4324/9781003306269-8

Outside of this book this myth is nicely called *Say's Law*. Overproduction cannot exist, at least not for long, according to Jean-Baptiste Say (1803).

The *Invisible Hand* and *Say's Law* are two examples of early economic myths. Today, they are not taken so literally. Numerous footnotes, limitations and extensions have been added. Yet the spirit of automatism still bears heavily upon economic theory. For instance:

> **The Trickle-Down Myth**
>
> *Prosperity trickles down automatically, so it's good that the rich get richer, because the poor will also benefit from that.*

This is what Simon Kuznets wrote.[67] He claimed that economic growth in capitalism leads to a decrease in inequality, both of income and wealth. He drew this conclusion based on research that was very limited, as he himself acknowledged. In his words: "The paper is perhaps 5 per cent empirical information and 95 per cent speculation." According to commentators, most literature on the *Trickle-Down Effect* is politically motivated.[68] Empirical data again and again show that in many cases the effect hardly ever applies or is even reversed, especially in the last decades: at a time of rapidly increasing prosperity, the poor hardly benefit and the inequality increases.[69]

Related to the *Trickle Down Effect*, there is the following myth.

> **The Rising Tide Myth**
>
> *Growth is a rising tide that lifts up all ships. Everyone benefits if the economy improves.*

The American president John F. Kennedy liked to say that, although he did not make it up himself (as many mistakenly believe). He and several of his successors, including Ronald Reagan, used the myth as an argument for lowering taxes. Obviously, the rich benefited considerably more than the poor, as Thomas Piketty proved on the basis of a great deal of data.[70] David Parker sighed: "We believe that a rising tide of economic growth should lift all boats, not just the super yachts."[71]

Various myths relate to economic equilibrium, which – you guessed it – automatically comes about and maintains itself. *Dynamically* maintains itself, which means that the forces in the economic system interact in such a way that their combination automatically pushes in the direction of stability. Nice huh? There is also:

> **The Perfect Market Myth**
>
> *The perfect market, which only knows perfect competition, is always in balance.*

Léon Walras introduced the myth in 1854. Such a perfect market is defined as a market that consists exclusively of companies that are too small to dominate the market. These companies are always fully informed and provide absolutely identical products – 'perfect competition'. As a result, their perfectly rational customers only have to pay attention to the prices so that they can maximize the 'utility' of their purchases.

Thoughts like these are tacitly based on a lot of naive assumptions. Have those who wrote or subscribed to it ever met real people? Probably not, because behind this illusion lies another myth.

The Homo Economicus Myth

All people have the following characteristics: (1) they are always one hundred percent selfish; (2) they always think and act one hundred percent rationally; and (3) they are fully informed at any time.

This myth is highly popular with various schools, including the neoclassical school. This is no surprise, because with these abstract creatures you can make fine models and calculate with them; this is much more difficult with real people. Joseph Stiglitz describes this *homo economicus* as "a calculating, rational, self-serving, and self-interested individual" with "no room for human empathy, public spiritedness, or altruism".

Homo economicus, an absurdly distorted simplification of *homo sapiens*, is actually applied as the prototype of humans in numerous economic models. Those who designed these models probably realized that simplification. The biggest problem is that those who then apply the models are generally unaware of this and all other simplifications. Just take:

The Efficient Market Myth

Bubbles, like those of house prices, cannot occur at all, since the free market automatically ensures that the prices are at the right level at all times.

I mentioned Eugene Fama, the source of this brilliant statement (which earned him the Nobel Prize in Economics in 2013), in Chapter 2 (page 16). This myth is widely known in economic circles as the *Efficient Market Hypothesis* (EMH). Actually, the EMH is nothing new, it is just a variant of the other myths that I showed you. Yet it is interesting, because it proves that even today, automatisms are still gloriously worshiped, despite the overwhelming evidence to the contrary. According to the EMH, there are no bubbles, they never existed and they cannot exist at all. The fact that in reality they have occurred and still do,[72] is apparently 'too bad for reality', as it is called.*

* The statement is, probably wrongly, attributed to philosopher Hegel, who would have said (but did not write down anywhere): "If reality does not fit the concept, too bad for reality." ("Wenn die Wirklichkeit nicht mit der Theorie übereinstimmt, desto schlimmer für die Wirklichkeit.")

There are still more automatism myths. Like ***Gossen's Second Law***,[73] which states, among other things: *Consumers always spend their income optimally*. This myth is still popular, despite repeated refutations.[74] Or the ***Law of the Comparative Benefit***, conceived two hundred years ago by David Ricardo,[75] who claimed that *unrestricted free trade between countries automatically leads to a balance in their trade*: no deficit or surplus. This law is considered by many economists to be one of the finest results of economic science;[76] by contrast, it is labeled by others to be "something of an article of faith",[77] firmly demolished on both theoretical and empirical grounds,[78] and still extremely popular.

Do you want some more fancy myths? Google on the ***Coase theorem***.[79] I leave it at that for now, I think you have the general idea.

7.2 More myths

It is shocking to see that quite a few economists believe that it is useful to consider their childlike myths as a reasonable approach to reality. Yet that is the case. According to the supporters of the neoclassical model, or the *School of Chicago*, the real market differs little from the perfect market, so that the theory can soundly be applied in practice, provided with a few corrections. According to them, deviations from perfection are limited and short-lived.

This neoclassical model is, and has been, the predominant economic view for many years, shaping policies with many governments and financial institutions.

The summary of all automatisms is: do nothing, then everything will be fine. Obviously this is an excuse, because those who propagate this actually mean: do nothing, then everything remains as it is; the rich remain rich (or become richer), the powerful remain powerful (or become even more powerful). The official term for this is: maintaining the status quo, the *established order*.

That is why influential governments are regarded as disturbing by economists from the dominant schools, since governments sometimes have the tendency to *do* something.

It will not surprise you that most influential economists and politicians have an income that is well above average. When they talk about the egoistic and calculating *homo economicus*, do they perhaps derive that image from the mirror?

There is another myth that has the task of strengthening the image that the current economy is good for everyone.

The Myth of the Blessings of Capitalism

Capitalism, with its free market which is unimpeded by governments, has brought us our present well-being.

Certainly, the free market produces a lot of prosperity, at least for the lucky ones who come to the top, or at least in the middle class in a western country. But many of the achievements of western society, such as women's suffrage, human rights, minimum wages, holidays, collective labor agreements, health insurance, press freedom and environmental policy, have been fought against capitalism by a self-conscious citizenry, partly through a democratic and far-reaching government formed by them.[80] So, these welfare aspects exist *despite and not thanks to the free market*. In addition to prosperity, the capitalist system has also brought a lot of misery and danger, as I will describe in detail in the next chapter.

Opposing the capitalist myth, there is the communist myth.

> **The Revolution Myth**
>
> *A violent upheaval will make it all right.*

This is the myth of Marxism,[81] and indeed of the French Revolution and many other violent revolutions. If you believe that the rich and powerful should give up their place, then you can hope to achieve that by upturning society, so that the upper ones will be the lowest and vice versa. Obviously, this is an illusion. You can imagine society as a pyramid, with some rich and powerful at the top and many powerless down-and-outers at the bottom. If you turn that upside down, many will arrive at the top. But, of course, it doesn't stay that way, because some of the former poverty strikers work themselves above the others while the rest collapses, and eventually you get the normal pyramid back, only with other rich and powerful at the top. The new rulers are often worse than the old ones: think of Napoleon, Stalin and Mao Zedong. I do not know of any example where a large-scale outburst of violence directly led to something good.

Do you want more myths? I can recommend *Economyths* by David Orrell: he created a collection just like I did.[82] Here I just add two other remarkable myths.

> **The Value Myth**
>
> *Value = money.*

This myth was already discussed in Chapter 3 (page 23). I will explain the second one here:

> **The Money Myth**
>
> *Economics is not about money.*

It may sound bizarre, but many macroeconomists believe that economics has little or nothing to do with money. For example, Mervyn King, leading macro-economist and governor of the *Bank of England* from 2003 to 2013, noted: "Most people believe that economics is about money. Yet most economists hold conversations in which the word 'money' appears hardly at all."

According to David Orrell, who quotes King, the reason why the central banks did not foresee the banking crisis of 2008 is that the banks simply do not appear in their models. Orrell calls this miraculous view "a gaping hole in the theory, a blind spot in the expert knowledge, like saying that the doctors haven't quite figured out the role of that red sticky stuff so are just going to ignore it for now, though it would be great if it stopped leaking out when people are being treated".

Others are amazed too. Alejandro Nadal calls money "without doubt the most important economic object" and writes: "The fact that money was exiled from the analysis of market processes as the discipline was being born and that this tradition has been maintained until our days is one of the most remarkable facts of economic theory."[83]

Ricardo Reis fully agrees. He is not the least: he was one of the youngest professors ever at Columbia University, before he accepted a professorship at the London School of Economics. Reis was harsh when he wrote, referring to the *Money Myth* and to the Great Recession of 2008[84]:

> "A (...) criticism of macroeconomics is that it ignores financial factors. Macroeconomists supposedly failed to anticipate the crisis because they were enamoured of models where financial markets and institutions were absent, as all financing was assumed to be efficient."[85]

"All financing assumed to be efficient." "Enamoured of models": *in love*. Do we need any more proof that we are not dealing with science here?

Sure. Look.

7.3 Copied methods

So far you have seen examples of different types of activities to create an economic science. I wrote about school formation, about pattern recognition (such as the presumption of cycles), and about myth creation. But new science developers have more options. You can also try to imitate successful methods from other scientific fields. But that does not always work well.

Science and mathematics

In his younger years, Isaac Newton, who would later realize the breakthrough of physics and astronomy as a science, devised a completely new branch of mathematics:

differential calculus.* For his physics theories, this form of mathematics proved absolutely indispensable. Together with other innovations, such as new instruments, the new mathematics gave a powerful impetus to the emergence of the first scientific discipline.

For the birth in the 18th century of the next major field of science, chemistry, mathematics was not of such great importance. In fact, in 1830 philosopher Auguste Comte wrote:

> "Every attempt to employ mathematical methods in the study of chemical questions must be considered profoundly irrational and contrary to the spirit of chemistry... if mathematical analysis should ever hold a prominent place in chemistry – an aberration which is happily almost impossible – it would occasion a rapid and widespread degeneration of that science."[86]

That all changed drastically in developments in the study of chemistry. But in biology, which broke through in the 19th century, there was a completely different branch of mathematics that played an indispensable role. This was *probability calculus*, which had a crucial significance for Mendel's theories about the inheritance of plant characteristics, which would later form the basis of genetics. Gregor Mendel, who announced the results of his years of experiments with the crossing of peas in 1865, received little recognition during his lifetime. Darwin did not know his work, so Mendel's contribution to biology forms a separate entrance to the biological success narrative.

Medicine and the social sciences benefit from yet another branch of mathematics: *statistics*.

You can draw two important lessons from all this.

The first lesson is that mathematics has often, but not always, made a crucial contribution to the emergence or progress of a specific field of science.

The second lesson is that it is crucial that every branch of science selects or develops its *own* methods and techniques. It is naïve and wrong to indiscriminately adopt the methods of other sciences, simply because they were successful there.

Physicalization, simplification

What did economics do? It attempted to match the successes of physics by copying its methods. This process is called *physicalization*. Attempts have even been made to literally take over the concept of 'gravity', in order to calculate trade flows on the basis of the economic size (the 'mass') of trading countries. Don't think that it was a primitive beginner's mistake in the 18th century or so: the idea was introduced by Isard in 1954, extended by Jan Tinbergen[87] and, with the necessary ad hoc adjustments, is still in use.[88]

* Newton himself called it the *Method of Fluxions*, see Newton (1671). More or less simultaneously, Leibniz also developed this branch of mathematics.

Naturally, such an approach means that, just like in physics, you provide quantities with abbreviations and build mathematical formulas with them. This task is mainly reserved for a special subsection, *econometrics*. In it, models are created in mathematical language. Such as: the *Arrow-Debreu* Model.

An example: Arrow-Debreu

A flagship of today's economy is the theory of the *Dynamic Stochastic General Equilibrium* (DSGE). Allow me to explain the terms. An *equilibrium* is *general* when supply and demand of all products and services are in balance with each other. It is *stochastically* balanced if it is resistant to random shocks and disturbances. And if it's a *dynamic* equilibrium, it will remain in balance over the course of time, even if various system variables change, such as labor input and national income, because they move sufficiently together under the influence of forces that keep each other in balance.

The DSGE principle therefore resembles what I defined in Chapter 4 (page 40) as the task of omniconomics: the *Broad Holloway*. At first glance, this similarity seems strong. The econometricians even went a step further by giving mathematical forms to the model.

A popular example is the *Arrow-Debreu Model.*[89] Its developers proved – strictly mathematically – that Léon Walras was right in 1854: in a perfect market, a state of equilibrium is always mathematically possible. To put it more accurately: Under certain economic assumptions, theoretical grounds require the existence of a composition of prices so that the total supply of goods and services is equal to the total demand for them. The necessary assumptions include perfect competition, i.e. a perfect market without government intervention: one of the myths already mentioned. With such unrealistic conditions, the model – which is a hit among modern econometricians – is effectively invalidated right from the start by its designers Arrow & Debreu. But that is not all.

A striking weakness of the *Arrow-Debreu Model* is that while it theoretically demonstrates that every perfect market must have a dynamic balance point, it does not say how stable this balance point is. The evidence does not exclude that a very small ('stochastic', i.e. random) disruption is too much for the stability, causing the system to get out of hand. Do you remember Figure 3.1 on page 21? On top of the mountain is a tiny stable area – the dimple – but only a small disturbance is needed to get out of that equilibrium area. When seeing this mountain top, with precipices on all sides, such a DSGE model suddenly matches much less with my Broad Holloway than before.

That is why this DSGE model raises a question. *What is the use* of mathematically proving that an idealized, far from reality system should theoretically be in a state of equilibrium if you: (a) do not show that it is a *stable* equilibrium; and (b) just *know* that the model is not realistic, not even a little? The real economic system has *not* been in equilibrium for centuries; on the contrary: it always strikes so seriously out of balance that disasters occur all the time. And there are many threats that can cause the system to explode completely in the not-too-distant future. *So why are you actually talking about such a model at all?*

Criticism from insiders

Fortunately, I can make an important side note on this whole narrative. Not all economists naively believe in DSGE-like mathematical models. And again, this shows how deeply economists disagree among themselves about leading ideas on the economy.

Let me illustrate this with a few quotes. I start with Joseph Stiglitz, who was mentioned before. He was chief economist at the World Bank and winner of the Nobel Prize for Economics. He wrote[90]:

> "I believe that most of the core constituents of the DSGE model are flawed – sufficiently badly flawed that they do not provide even a good starting point for constructing a good macroeconomic model.
>
> These include (a) the theory of consumption; (b) the theory of expectations – rational expectations and common knowledge; (c) the theory of investment; (d) the use of the representative agent model (…); (e) the theory of financial markets and money; (f) aggregation (…); (g) shocks – the sources of perturbation to the economy; and (h) the theory of adjustment to shocks."

In short, according to Stiglitz, this relates to just about *every aspect* of the favorite model of contemporary economic thinking.

I also quote Olivier Blanchard, chief economist of the International Monetary Fund (IMF) from 2008 to 2015. He wrote about DSGE:[91]

> "Current DSGE models (…) are based on not just simplifying assumptions, as any model must be, but assumptions profoundly at odds with what we know about consumers and firms."
>
> "The benchmark New Keynesian model, from which DSGEs derive their bone structure, is composed of three equations (…). At least the first two are badly flawed descriptions of reality."
>
> "The models come (…) with a very large number of parameters to estimate, so that classical estimation of the full set is unfeasible. Thus, a number of parameters are set a priori, through 'calibration'. This approach (…) is highly suspicious."

In short: if economics is ever to become a serious and realistic discipline, it will first have to devise its own appropriate methods, instead of blindly copying methods from other disciplines.

7.4 Oversimplification

"Simplifying assumptions, as any model must," I just quoted Olivier Blanchard. He is absolutely right. *Every* model is a simplified representation of reality, and that cannot be otherwise: a model that would describe reality completely and correctly, would be at least as large as the universe. Physical, chemical and biological laws also leave out aspects of reality.

The only question is: how far can you go with these simplifications? At least not so far that the results no longer have anything to do with reality, causing the results to be misleading and therefore downright devastating for all of humanity and for life on earth. If you do that, you are oversimplifying, which is one of the characteristics of protoscience that I mentioned in Chapter 6 on page 59.

The long list of automatisms and other economic myths are all ways of oversimplifying economic reality. The *homo economicus*; the idea that the economy has nothing to do with money; and narratives such as that of the rising tide: they all testify to an oversimplification that goes so far that only a caricature, a cartoon, of reality remains.

Figure 7.1 illustrates this. The figure highlights only one ripping effect of contemporary economic reality: the absurdly growing inequality in income, wealth, power and prosperity between people and between individuals and multinational corporations.

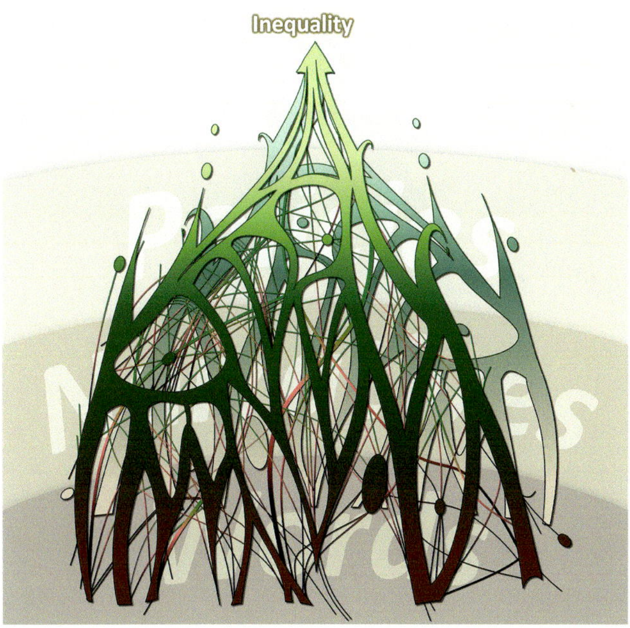

FIGURE 7.1 Reality is extremely complex. This image symbolically depicts the roots of only one societal problem: the rapidly growing inequality.

In the predominant economic models, the highly complex interactions, which I have symbolically drawn in Figure 7.1, have been extremely simplified to the simple straight lines of Figure 7.2. Next, part of this simplified caricature is split in two: the blue colored central part symbolizes the economic domain, i.e., things whose value can be expressed in money, while all the grey areas around it symbolize what the economy considers to be externalities, i.e. freely available and destructible. While Figure 7.1 shows the roots of growing inequality, Figure 7.2 emphasizes 'the' wealth that is growing: it just depends on what you want to see.

The gravest damage then arises when theoretical economists, followed by many politicians and administrators, forgot (or never knew) that Figure 7.2 is a gross imitation of Figure 7.1, after which they take the conclusions from the childlike economic models and theories, apply them to reality and base their political and economic governmental or enterprise policies on them. No, it is no wonder that the economic system explodes from time to time, just as the climate does. Nor is it surprising that the growing inequality is tearing apart societies, just as has happened repeatedly over the past few thousand years.

FIGURE 7.2 The economy simplifies complex reality into a small number of linear objects and interactions. Blue = the domain of the economy; gray = externalities.

Physicalized laws

Another aspect of the physicalization of economics is the widely held, but usually tacitly professed, belief that the laws of economics have the same status as the laws of physics, which would mean that they are necessary and universal. Łukasz Hardt writes about this in 2021[92]:

> "Although many economists would agree that laws they discover do not have the status of laws of nature, they still have many problems in rejecting the worldview which sees the world as being governed by such [economic] laws."

But even that is not true, because unlike the laws of physics, which were *discovered* by humans, the laws of economics were *created* by humans, just like legal laws. They are a *choice*.

Such a misconception of economic laws creates the illusion that the economic system *cannot be anything other* than what it is, thus rendering many choices invisible from the outset.

The illusion of the necessity of the current economic laws is a *myth*:

The Myth of the Universal Economic System

The fundaments of the economic system are necessary and cannot be changed. The economic laws that describe these fundaments are, like laws of physics, universal.

It is extremely important to realize that this is indeed a myth. For if you do not understand and recognize this, you limit your thoughts, your imagination and your creativity, and from a tunnel vision you are at best only able to think of superficial improvements to the prevailing economic system. And as I wrote in the opening chapters: *That does not go deep enough.*

7.5 Lack of replicable experiments

'Too bad for reality', I wrote a few pages before. This concerned the countless situations where economic models claim something else than what we see in the real world. This is similar to the situation in the Middle Ages, about which I wrote in Chapter 5:

> These claims were not systematically tested. This should not come as a surprise, as they not only lacked not only the right equipment, but also the interest. Testing hypotheses was primarily done by the head and not with the eyes: intuition and a sense of logic, rather than observations, dictated the evaluation

Indeed, there is little interest in the empirical testing of economic models. But to be honest, the field does not have a very wide range of possibilities for conducting experiments. Where in other fields there is ample opportunity to replicate an experiment numerous times, with or without variations, economics has little at its disposal.

You can distinguish a number of types of experiments that economics tends to apply.

In the first place, there are investigations into the behavior of individual people: in their role either as a responsible citizen, or a consumer, or a professional in a working environment. In fact, this is more about psychological or social experiments than it is about economic ones. A disadvantage of such experiments is that they are particularly difficult to reproduce. In recent years, many old studies have been repeated, often showing that the previous results were not confirmed,[93] including experiments that were explicitly qualified as economic in nature.[94] Another drawback is that the experiments are usually small-scale and say little about the behavior of much larger groups of people, while that is what the study of global economics needs. Further on in the book I will show that, for this reason, such experiments are not really *economic* experiments at all.

Another group of experiments is truly of an economic nature, although usually they are not intended to be scientific experiments. These are locally established economies, often with their own currency or exchange system, usually with alternative visions on the relationships between people, companies and nature. Permaculture, organic farming, local products and food parks are among the key words of this approach. The strength, and at the same time the weakness, of these experiments, is that they take place on a scale where the distance between people is small. Often everyone knows each other, or in any case even the highest 'degree of separation'* is low. This locality with the resulting sense of togetherness is an important factor in success. The weakness, seen from a macroeconomic perspective, is that there is a big chance that these experiments fail as soon as you try to scale them up to the size of a large country or even the entire world.

Also, existing companies, governments and other organizations are studied. Their characteristics and behavior are compared with the results they achieve, in a positive (financial profit, societal contribution) or in a negative sense (damage and pollution, nuisance and danger, use of scarce resources). The scale of such investigations varies; the replicability is usually only small. The biggest weakness of such studies is that all these organizations (of course) operate within the existing economic system, so they teach us little or nothing about radical innovations in that system itself.

Then, there is a category of experiments that is really large-scale. The introduction of the euro is an example of this, as is Brexit, the departure of the United Kingdom from the European Union, which, by the way, is itself such an experiment.

* The number of steps that you have to take from person to person in order to get from someone to someone else, such that at each step the persons involved know each other directly.

The examples show that these far-reaching changes are not intended as scientific experiments but as desirable, permanent *transitions* of a sizeable system, based on the conviction that the 'experiment' will succeed.

Not only are such experiments rarely, if ever, replicable, they also carry a huge risk. This is evident from other examples of such experiments: the introduction of communism in the Soviet Union, first designed under the direction of Lenin and then of Stalin; and its Chinese variant, introduced by Mao Zedong. Both experiments were disastrous and caused the death or other misery for hundreds of millions of people. Not only are such experiments not replicable; they are ethically unacceptable.

And so, economics is struggling with an ugly problem. Is the lack of opportunities for relevant large-scale and replicable experiments the cause of the lack of interest in empirical testing? Or does this stem from the self-interest of the often wealthy researchers and administrators, who may be more interested in maintaining the established order than in new experiments? Perhaps both explanations contain a germ of truth. In any case, there is no tradition in economics of replicating experiments, as McCullough and colleagues wrote, based on an analysis of a long series of publications on economic experiments, with deadly precision in 2006[95]:

> "In 1989, Stanley Pons and Martin Fleischman claimed to have produced 'cold fusion' which, if true, would have led to an almost limitless supply of cheap energy. Upon the announcement, researchers around the world began attempting to replicate this already-famous result. Before long, MIT, Cal Tech, Harwell, and others reported failures to confirm. The scientific method had revealed the claim of cold fusion to be false.
>
> If Pons and Fleischman had published their cold fusion result in an economics journal, the world would still be awaiting lower utility bills. Distinct from most sciences, **economics has not fully embraced the scientific method**; in particular, there is no tradition of replication in economics. Results published in economics journals are accepted at face value and rarely subjected to the independent verification that is the cornerstone of the scientific method."

The bold text (I added the bold) shows again how economics is not a mature science. It's for a reason, that the lack of replicable experiments – indeed, the lack of need for them – is one of the characteristic features of protoscience that I listed in Chapter 6 (page 59).

7.6 Economics is a myth

Lack of replicable experiments. Oversimplification. Copied methods. Automatisms and other myths. Naïve, assumed patterns. A large number of schools and impetus words. Those are all characteristics of protoscience that typify the field of economics.

Will economics be able to overcome all these limitations in the short term? Probably not. In 2018, Blanchard was absolutely crushing about a solution to the eternal struggles between the economic schools:[96]

> "The pursuit of a widely accepted analytical macroeconomic core, in which to locate discussions and extensions, may be a pipe dream."

If Blanchard's conjecture is correct, macroeconomics will *never* be able to become more than a protoscience. I think he is right, because a mature science can only emerge if the discipline is defined more broadly and an omniconomic science is developed.

Actually, this means that the current discipline of 'macroeconomics' itself is a myth. Not only for the reasons mentioned, but also for another. To demonstrate this, I first show how the already mentioned Joseph Stiglitz defines the field. The quote comes from the fourth edition of his authoritative university textbook on economics, an edition that was co-authored by Carl Walsh.

Economics is the discipline that[97]:

> *"… studies how individuals, firms, government, and other organizations within our society make choices, and how these choices determine society's use of its resources. Why did consumers choose to buy small, energy-efficient cars in the 1970s and large sports utility vehicles in the 1990s? What determines how many individuals work in health care industries and how many work in the computer industry? Why did the income gap between rich and poor rise in the 1980s? To understand how choices are made and how these choices affect the use of society's resources, we must examine five concepts that play an important role: trade-offs, incentives, exchange, information, and distribution.*
>
> 1. *Choice involves* ***trade-offs*** *– deciding to spend more on one thing leaves less to spend on something else; devoting more time to studying economics leaves less time to study physics.*
> 2. *In making choices, individuals respond to* ***incentives****. If the price of Zen MP3 players falls relative to the price of iPods, there is a greater incentive to buy a Zen. If the salaries for engineers rise relative to the salaries of people with an MBA, there is an increased incentive to choose to study for an engineering degree rather than a business degree.*
> 3. *When we* ***exchange*** *with others, our range of choices becomes larger.*
> 4. *Making intelligent choices requires that we have, and utilize,* ***information****.*
> 5. *Finally, the choices we make – about how much education to get, what occupation to enter, and what goods and service to buy – determine the* ***distribution*** *of wealth and income in our society.*

> *These five concepts define the core ideas that are critical to understanding economics. They also guide the way economists think about issues and problems. Learning to 'think like an economist' means learning how to discover the trade-offs and incentives faced, the implications of exchange, the role of information, and the consequences for distribution."*

This widely accepted definition is about distribution, about trade, about production and consumption. At the same time, many essential aspects are missing, and that makes the definition *fundamentally wrong*.

Compare it with a business plan for a company. What would you think if you read everything about the customers and the products or services that the company provides to them in that plan; but nothing at all about the suppliers, the purchase of raw materials, the stock management, a comparison between the expenses and the income, and the physical location? Such a plan would be seriously incomplete. If you went to a bank with such a plan to get financing, you would get a big 'inadequate' – and no money.

Planet, Ltd.

Now consider – just for a moment – the world economy as a company. Let's call it *Planet, Ltd.* In that case, the major supplier is obviously nature, both living and lifeless, including the Sun. The acquisition of raw materials is provided by agriculture and mining. Inventory management does not only relate to semi-finished or finished products, but also to raw materials and energy, and therefore also to the future expectation of the security of supply, that is to say, the preservation of nature.

The comparison between income and expenditure relates on the one hand to the financial and other revenues from production, and on the other hand to expenditures due to extraction from nature, including damage, loss and risks. The profit should be defined as the difference between them. But, in reality, the 'profit' that countries book annually, defined as the gross domestic product (GDP), is calculated exclusively on the basis of the revenue side and therefore by definition gives a wrong outcome. GDP is an *impetus word*. (See Appendix 3 for an explanation.) The same applies to the sum of all national GDPs, the annual profit of the entire world: the Gross World Product (GWP).

Finally: as far as the physical location is concerned: for the world economy, this is obviously planet Earth. Where a normal company makes an Environmental Impact Assessment (EIA) for each investment plan (legally required!), such an EIA is missing in the 'business plan' as Stiglitz and Walsh defined it.

A definition of 'economics' that completely ignores such essential aspects is largely incomplete. Can you show even more clearly that economics wrongly regards nature as an externality? All this proves that the narrative called 'economics' is not right. It's a myth.

> **The Myth of Economics**
>
> *Economics is an independent discipline with the status of science, with a clearly defined study area.*

7.7 Conclusion: economics is a protoscience

There are quite a few economists who are very critical of their field. But when it comes to solutions of the shortcomings, some go further than others.

I already mentioned the criticism of Stiglitz, Reis, Blanchard and others on the popular DSGE model and its mathematical elaborations. But, according to their publications, these economists have set their hopes on developing stronger, more realistic models based on the same fundamental concepts. Their paradigms, the words in which they try to tell new narratives, remain largely the same.

I conclude that the field of 'economics' meets at least six of the seven criteria of a 'protoscience'. According to the definition, a discipline is a protoscience if it "possesses all or most of the seven characteristics". It has thus been demonstrated that economics is, according to the definition I have given, a protoscience. Nevertheless, it is important to investigate the seventh and final characteristic: no or little predictive power or other form of 'success'. In order to investigate this, first this 'success' must be defined in detail.

Recommended further reading (for details, consult the online Bibliography)

Book: Orrell, David (edition 2017): *Economyths. 11 ways economics gets it wrong.*
Book chapter: Reis, Ricardo (2018): *Is something really wrong with macroeconomics?*
Book chapter: Hardt, Łukasz (2021): *An Essay on Humble Economics.*
Journal article: McCullough, B.D., K.A. McGeary, T.D. Harrison (2006): *Lessons from the JMCB Archive.*

8

THE UNSUSTAINABLE SUCCESS OF ECONOMICS

8.1 Scientific success

A science can be successful in different ways. By the way, I am not talking about the success of individual scientists: fame, status, awards. The success I am talking about here concerns entire fields of science, from physics to neuropsychology. And economics?

Such a field of science can be successful in different ways. I distinguish three types of success, each of which consists of three components. You may interpret the overview as a definition.

Scientific success consists of:

Intellectual output:
- Order • Understanding • Theoretical edifice

Societal output:
- Predictive power • Technology • Contributions to society

Emotional output:
- Amazement, beauty • Insight into who we are • New questions

DOI: 10.4324/9781003306269-9

Here, I limit myself to a few brief explanations. In later chapters, I will go much deeper, as a starting point for the formation of an omniconomic science.

Order is the bringing of order out of chaos, in the form of pattern recognition; of fixed relations of the type 'If A, then always B'; of rhythm, rhyme or repetition.

Ideally, order leads to ***understanding***. This is a very complex concept: later on, I will devote a whole chapter to understand 'understanding'. Let me announce right here that 'understanding' has nothing, absolutely nothing, to do with 'truth'. If that makes you curious or even indignant, then quickly fly on over to Chapter 13.

If this understanding is extraordinarily successful, it results in a grand, widely accepted ***theoretical edifice***, consisting of facts, laws, hypotheses and theories, all of which continue to be carefully tested by experiment and observation. In this structure, enormous amounts of knowledge are constructively linked to each other, so that they have a logical coherence. I will describe how this works in Chapter 15.

Predictive power is what many people think of first, when it comes to scientific success.[98] Here is a pleasant example.[99]

Good news: Apophis will not strike this century

March 26 2021

We already knew that Apophis, the asteroid named after the Egyptian god of chaos and death, will pass close to Earth in 2029 and again in 2036 without hitting us. But more sophisticated observations and calculations have now shown that even the close passage in 2068 will not result in a cosmic collision that could destroy Earth's civilization worldwide. As far as Apophis is concerned, we are safe at least until the end of the 21st century, according to NASA.

Nice to know, though. This prediction is made possible by another form of success: ***technology***, including satellites and TVs. That kind of success need not be confined to high-tech electronic equipment; it also includes seed breeding, vaccinations, hypnotherapy and business assessments, provided they are firmly evidence-based, with which the various sciences make their ***contributions to society***.

And then there is the emotional value of science, because it leads to fascination and ***amazement***. About bizarre discoveries; about unimaginably large or, on the contrary, smaller-than-microscopic worlds; about infinities. Amazement at the ***beauty*** and *elegance* of the grand edifices constructed by science. About the beauty of reality itself. With as building blocks the new words with which we tell new stories: about the origin of the Universe from a Big Bang, or about the origin of humanity via the evolutionary descent from ape-like creatures: discoveries that give us ***insight into who we are***. These new insights then always inspire ***new questions*** – see Chapter 15 – so that our curiosity will always be stimulated and the human adventure never ends.

Economics

The definition of 'success' in science is intended to prepare for an examination of the degree of success of the field of economics that follows right now.

I can be brief about some of the forms of success. When it comes to ***order*** and ***understanding***, the vast economic mythology makes it immediately clear that the discipline does not score here.

The staggering multiplicity of schools makes a grand, widely accepted and coherent ***theoretical edifice*** impossible.

The staggering lack of ***predictive power*** became apparent in Chapter 2, which on page 14 discussed, among other things, the unexpected collapse of the global economy around 2008, the Great Recession. (That the economic crisis caused by the 2020 pandemic was not predicted is less culpable, as it had a cause outside the economy.) Try to imagine[100]:

> **Good news: Recession will not hit this century**
>
> *March 26 2021*
>
> *We already knew that the Recession will pass close to the Earth's economy in 2029 and again in 2036 without hitting us. But more sophisticated observations and calculations have now shown that even the close passage in 2068 will not result in an economic collapse that could devastate the Earth's civilization worldwide. As far as the Recession is concerned, we are safe at least until the end of the 21st century, according to the World Bank.*

Unthinkable, obviously. Absurd. *Yes, but* – people sometimes argue angrily when I talk to them about this. *Yes, but that is not fair: the economy cannot be that predictable, because it is about people, and people are fundamentally capricious and unpredictable.* My response to that is: economics – or rather: omniconomics – is not about individual people but about something else entirely. To tell how and why, I need several other chapters as preparation.

As for the emotional successes of economics, I would like to say that it does not lead to ***amazement*** but rather to ***bewilderment***. In me and others.

First, in this chapter, I will examine ***technology***, and thus the ***contributions to society***, of the discipline of economics.

Then, in the next chapter, I will look at the significance of economics for ***insight into who we are***.

8.2 Growth

For several centuries, the number of people in the world has been growing rapidly. Figure 8.1 shows this growth from the year 1500 onwards. The graph rises to over

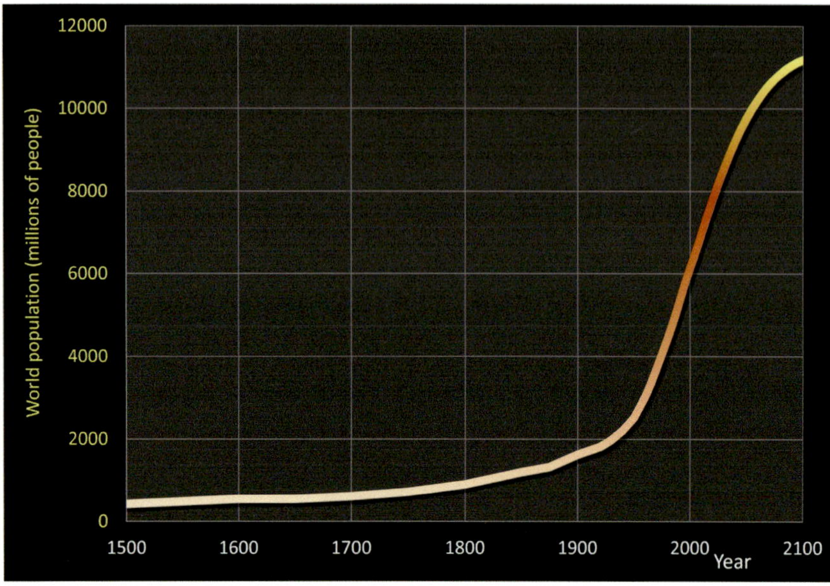

FIGURE 8.1 Size of the world population from 1500 to 2100; the rightmost part is an expectation.[101]

11 billion people by the year 2100, in line with the latest projections from the United Nations Department of Economic and Social Affairs (UNDESA).

What the graph shows is an extremely powerful exponential growth beyond the year 2000. In fact, the rate of growth in the 20th century was more than exponential, because the doubling time, which is constant in purely exponential growth, was actually getting shorter and shorter – *despite world wars, famines, epidemics and Stalinist and Maoist humanitarian catastrophes* – until the doubling time was only 35 years two-thirds of the way into the 20th century. That corresponds to 2% growth per year.

That rate of growth is staggering. It means an eightfold increase in the world's population per century. If growth continued at the same rate, in the year **3540** our entire ***planet*** from crust to core would consist of solid human tissue, in **5633** the entire ***Milky Way Galaxy*** with its hundred billion stars and planets, and in **6761**, when humanity is only one and a half percent older than today, the entire observable ***universe***... If you can't believe that: the calculation is in Appendix 5. The absurd result shows: the growth of the number of people is going to stop one day. That is certain. And sooner than in 3540.

Much sooner even, because otherwise the Earth will be damaged beyond repair. The first signs of deceleration are fortunately already visible in Figure 8.1, although this is an expectation, not a certainty.

Will the deceleration be on time and sufficient to prevent disasters? The answer to this crucial question depends not only on the speed at which humanity grows. Just as decisive is the growth of average *production per person* and the resulting demand

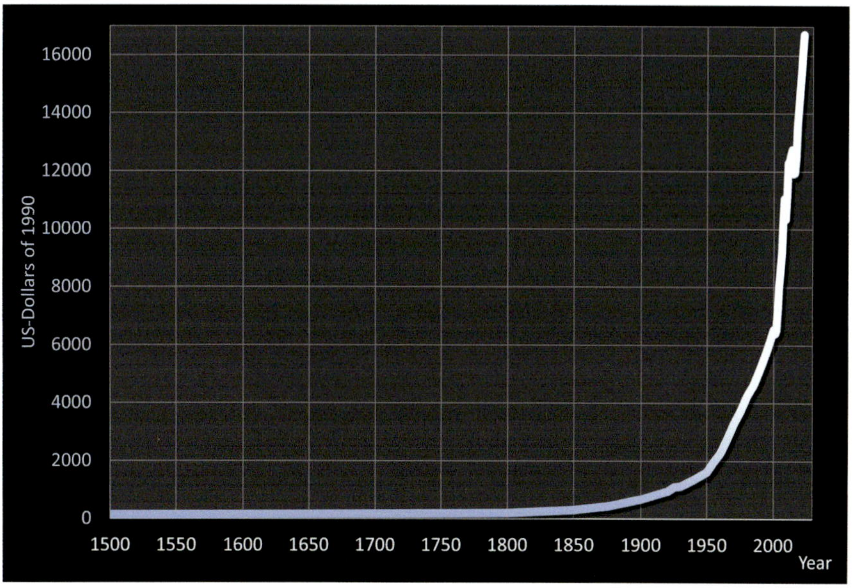

FIGURE 8.2 Gross World Product (GWP) per person.[102]

on raw materials, energy and the environment. The graph of production development is shown in Figure 8.2. The vertical axis shows the *Gross World Product (GWP)* per person. This is a measure of the average material prosperity, calculated over all people living at a given moment: rich and poor combined. The values have all been converted to 1990 US dollars, i.e. they have been corrected for inflation and can therefore be compared with each other.

To see what this means for the global production of all people together, the growth of humanity in Figure 8.1 must be combined with the growth of GDP per person in Figure 8.2. This time this is not done by *adding up* the values for each year, as was the case in Chapter 6 with the graphs in Figure 6.5 (page 68). The total prosperity is obtained by *multiplying* the average production per person by the number of people. The result is shown in Figure 8.3, which shows the outcome from the year 1900 onwards. In this figure, both the number of people and the GDP per capita in 1900 are set to '1', which means that the total GDP in that year is also 1: $1 \times 1 = 1$.

The graph shows that world production has increased more than 120 times since 1900. Since the beginning of the industrial revolution, around 1750, it has even increased *more than a thousandfold*.

Growth is necessary?

The strong growth of GDP can be seen as a great success of the economy, which has created enormous prosperity for a part of humanity. At the same time, we can see that, precisely because of this success, 'growth' has become such a permanent

the dizzying rise in prosperity, hundreds of millions of people still live in abject and unacceptable poverty. In addition, poverty is an important 'driver' of population growth, because as long as pensions are inadequate, children are the main source of retirement income; and where health care is inadequate or unaffordable, child mortality is high, so that those who want to be assured of having children in their old age, ensure that many are born. Wherever prosperity increases, population growth immediately drops sharply.[110] This means that, from the point of view of both humanity and solidarity, and in order to slow down population growth (which is in everyone's interest, including that of the rich), the poor need to catch up in their level of prosperity. This requires global economic growth, unless the economic growth of the poor can be compensated by a modest contraction elsewhere in the world. Which implies redistribution of wealth, but according to Joseph Stiglitz that is not realistic, and so he "aims for general growth" anyway.[111]

Those are a lot of reasons. You might say: in reality, growth might be necessary because of a *combination* of all those hypotheses. But the quoted sources don't show this, because they all claim that 'their' reason is 'the' reason. All in all, one gets the impression that something else is going on. Economists, politicians, CEOs and citizens just *want* growth, and that is the real, underlying reason. Politicians want growth to keep their voters happy. CEOs want growth for their companies because they see business as a game they want to win. Citizens want growth because they quickly get used to increases in prosperity, which become 'normal', after which dissatisfaction sets in that can only be cushioned by more increases of wealth.

Growth is therefore not so much a necessity as a goal in itself. I am not the only one who thinks so, for others wrote: "For most societies, economic growth is a fundamental goal."[112] And: "Economic growth is the grand objective. It is the aim of economic policy as a whole."[113] In fact, it is a myth:

> **The Economic Growth Myth**
>
> *We are not doing well when the economy is not growing.*

That growth is necessary to prevent economic misery is therefore disputed by various authors.[114] That is to say: in the *current economic system*, a lack of growth does indeed lead to recession or worse. But we can change that system, according to these authors: growth is a choice, not a necessity.

Dematerialization

From the 1970s onwards, publications appeared indicating that unlimited economic growth would eventually lead to catastrophe. The most famous example is the report *Limits to Growth* by the Meadows team,[115] which I will discuss in detail later in this book. The report was a smash hit, with both scientists and the general public.

But the vast majority of economists did not see continued growth as a problem at all. Meadows' disaster scenarios were widely vilified. Robert Solow wrote[116]: "I think that the Doomsday Models are bad science and therefore bad guides to public policy." After all, the free market would solve any problems *automatically!* For example, by learning to use raw materials more and more efficiently. As a result, according to Solow, "the overshoot and collapse are no longer the inevitable trajectory of the world system, and the typical assumption-conclusion of the Doomsday Model falls by the wayside."

Yeah: a new automatism myth was conceived and born.

The increase in the efficiency of resource use expected by Solow, which he and many other economists believe can be sustained indefinitely, is called *dematerialization*, and also *decoupling*: economic growth is increasingly decoupled from material growth, or so the assumption goes. This can be achieved, for example, by using new materials and smart technologies, including miniaturization (micro- and nanotechnology), and also by reusing and recycling, or, more generally, by following the circular economy principles. Furthermore, economic growth takes place partly in the form of new digital products, which do not necessarily lead to material growth, according to Solow and his colleagues.

So, the key question is: are there already signs that this dematerialization is successful? This has been investigated several times.[117] It was found that there is some, unfortunately modest, decoupling, but that this hardly leads to an actual reduction of the growth in the use of raw materials and energy, and certainly not to a reduction of the damage to the environment. Details on this can be found in Appendix 6. The conclusion is: dematerialization is a nice theory, which in the case of 100% decoupling would allow the economy to grow indefinitely while material and environmental use stay within healthy boundaries. But for now, all indications are that the decoupling will not even come close to 100%, with the result that an infinitely sustained economic growth will eventually lead to a situation in which our use of materials exceeds every physical limit.

Economic growth *will* stop, that is for sure: gently (if we discover how to stop in a pleasant way) or perforce (if we don't).[118] Jeremy Lent equates 'green growth' with 'magical thinking',[119] and Herman Daly sighed: "Sustainable growth is an oxymoron",[120] an internal contradiction, similar to terms such as 'the only choice', 'liquefied gas', 'compelling advice', 'working holiday'.

In my words: sustainable growth is a myth. 'Sustainable' and 'growth' are not compatible, at least not in the long term. *If, in the current economic system, growth cannot be missed, it logically follows that the system must be replaced by one that can.* If we are unable to do that, our global civilization is doomed.

Benefits and charges

The occupation of the Earth by humans was completed on a Tuesday.

If you always spend more than you receive, you will eventually run out of money and go bankrupt: this is a well-known financial and economic fact. You would expect that financial economists would understand better than anyone that we should not use up more nature than can be regrown. Funnily enough, that is not quite the case, because for years the planetary burden – the *ecological footprint* – has been greater than the benefits – the *biocapacity*.

By 1950, the global footprint had risen to *half* the biocapacity of the planet. That achievement was doubled in just 21 years, 1971 being the year when the combined human footprint covered 100% of the biocapacity. Everything was now for humankind; nothing was left for nature. The first *Earth Overshoot Day* (the day when the ecological 'year's budget' ran out before the end of the year) was on December 21 of that year. That was a Tuesday.

Since then, Earth Overshoot Day has been on an earlier date in almost every subsequent year, as Figure 8.4 shows. Where this is not the case, you can identify an economic crisis, such as the Great Recession around 2008 and the coronavirus crisis of 2020.

And then? When humankind had taken up the entire biocapacity, on that memorable day in December 1971? Surely, then the growth of the footprint stopped, because you can't take more than *everything*. Logical, isn't it?

If only it were true. The growth just went on. This means that from that moment on, humans take more than nature can provide. Locally, this had already happened much earlier, because the demands on nature naturally varied from place to place. But from 1971 onwards, the *global sum* has been negative. Since 1971, the entire Earth has been overexploited on balance.

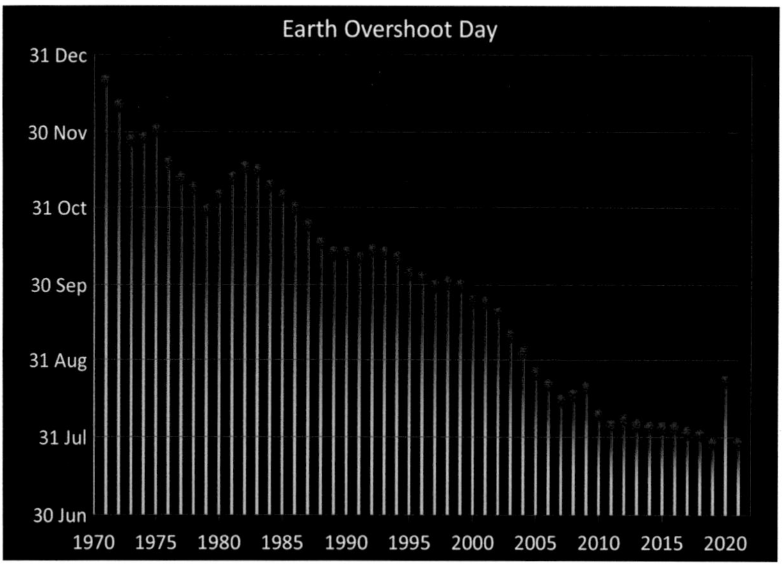

FIGURE 8.4 Earth overshoot day.
Source: Overshootday.org (2021).

With unpleasant consequences. Ancient forests are ploughed into fields. Fertile soil is being exhausted by intensive agriculture. Soil is worn away by erosion, after the roots of trees and plants have disappeared, causing valuable soil materials to be blown or washed away. Pastures are turned into cities, i.e. into stone plains. Monoculture dominates agriculture, exposing it to pests. Rivers are straightened for the benefit of shipping, which makes them flood faster. Corals at the bottom of the oceans die and crumble. Once-fertile areas turn into deserts, where dust storms blow for thousands of kilometers. Plant and animal species are dying out at a rate the Earth has not seen in 66 million years. The climate is changing, the temperature goes up, the sea is rising.

We are going bankrupt, we cannot keep this up.

8.3 Profit

Growth is not the only addiction of the global economy. *Profit* is another. Here is a bizarre starter.

The Restasis patent

The Saint Regis Mohawk Indians live in a reservation on the border of the American state of New York and the Canadian province of Ontario, separated from Canada by the Saint Lawrence River. They mainly live there from the income of their casinos. In mid-2017 these native Americans were made happy with a beautiful gift from a pharmaceutical company called **Allergan**.[121] They received the patent on *restasis*, a medicine against dry eyes, for free. That was not some irrelevant medicine, because it yielded Allergan a turnover of $1.5 billion every year. The Mohawks did not have to pay anything for their gift. In fact, they even got money for it, because Allergan immediately leased back the patent for an amount of $13.75 million for starters, plus $15 million annually.

Strange course of action, isn't it? To just give the patent away. Allergan achieved 10% of its turnover and even 15% of its profit from restasis. That is no wonder, because the popular drug was expensive: in 2008 it cost a mere $117 per box of thirty portions (active substance concentration: 0.05%), but the producing company happily used its exclusive right to sell and increased the price in steps to $274 per box in 2017. That meant the drug was the number two cash cow for the company: at number one was *botox*.

The patent on *restasis* would officially continue until 2024. But it was doubtful whether that would really happen, because the validity of the patent was disputed by competitors, including **Teva Pharmaceuticals**, **Mylan** and **Akorn**. They filed a lawsuit in 2015 to reverse the patent, so they could bring a much cheaper generic drug with the same active substance onto the market.

Now, the Saint Regis Mohawks reserve is a sovereign nation within the United States, based on a 1940 Supreme Court ruling. In their own words, the reserve is called *Mohawk Nation at Akwesasne*, a 21-square-mile (54-square-kilometer) nation with some 3300 citizens.

That fact gave the Texan law firm Shore Chan DePumpo an idea. Following their advice, **Allergan Pharmaceuticals**, who saw it coming to lose the lawsuit, donated the patent on *restasis* to the Mohawk nation. Because of the sovereign status of the country, American judges had no legal authority there!

That was, of course, a wonderful idea. Immediately, the Indians invited other companies to donate their endangered patents to them as well: they hoped that the patent earnings would soon surpass those from the casinos. In August 2017 computer producer **SRC Labs** came in, which also transferred patents and re-rented them, after which the Indian leaders immediately sued **Microsoft** and **Amazon** for infringements of 'their' patents.

This absurd story really happened. But it ended different than **Allergan** hoped. In October 2017, a judge put a stop to it by lifting the patent on *restasis* immediately. The judge spoke in his verdict of 'eyewash', which means 'eye water' but also 'nonsense'. In spite of this, the principle of the method was not prohibited, so for the Mohawks there is still hope for great extra income.

Is the *restasis* case misconduct? Everyone can judge for themselves; **Allergan's** appearance will not be called decent by many. This was also apparent from comments from American parliamentarians and media: the friendliest assessment was 'sham'.

At first glance, the story about *restasis* seems like a crazy exception. But there are many more examples of this kind, together showing that our economic world is seriously ill. The disease is called: *profit-seeking*.

Profit is *(according to my Dutch dictionary):*

1. Benefit obtained; proceeds received in excess of expenses incurred; what is gained.
2. Winning, ≈ victory, triumph.[122]

Profit is *(according to an English Dictionary):*

1. Excess of revenues over outlay and expenses in a business enterprise.
2. The monetary gain derived from a transaction.
3. Income derived from property or an investment.
4. The income or reward accruing to a successful entrepreneur and held to be the motivating factor of all economic activity in a capitalist economy.
5. Gain, benefit, or advantage.[123]

Gain is

1. To acquire (something desirable), obtain.
2. To win in a competition. (…)

The meanings of the word 'profit' fall roughly into two categories. The first is about the difference between income and expenditure, for example in the case of a transaction or an investment, and therefore about a healthy way of doing business. Most descriptions leave open whether this necessarily concerns profit in the form of money. The second meaning is: a victory in a game, a sport, a competition, a battle. The triumph, the feeling of superiority.

Is it the first or the second meaning, where Collins' dictionary states that profit is the 'reward that accrues to a successful entrepreneur'? Remarkably, it immediately adds that, in a capitalist system (i.e. in ours), profit is considered 'the' motivation for all economic activity. That is quite a statement. So, according to the editors of 'Collins', such things as business continuity, meaningful contributions to society and care for staff, customers and the environment do not really count: the 'real' entrepreneur goes for the glory!

The pharmaceutical industry

After a guest appearance by the drug industry in the *Restasis* story, it is now coming more fully on the stage, playing various roles in a number of unlikely but true stories. Together, these show that there is more to it than just a few isolated incidents.

As you read, it is a fun game to examine with each of the stories, which of the two types of 'profit' is most applicable there: sound business or triumph & superiority.

In India, two vaccines against pneumonia are on the market. **Pfizer's** *Prevnar13* costs €50 per child. **GlaxoSmithKline (GSK)** charges only €30 per child for *synflorix*, but even that is still completely unaffordable for hundreds of millions of Indian parents. India's **Serum Institute** is able to produce a vaccine that costs less than €5. But that is not allowed, because of existing patents on the more expensive drugs. According to Doctor Leena Menghaney of Médecins Sans Frontières, **Pfizer** and **GSK** have already made €30 billion profit from their vaccines. But in India, two hundred thousand children die every year from pneumonia.[124]

In 2015, the start-up company **Turing Pharmaceuticals** bought the drug *dataprim*, which is supposed to protect HIV patients and pregnant women from infection, for US$55 million. The drug costs 40 cents a pill in the Netherlands, but in the United States the price was raised overnight from $13.75 to $750. Per pill. The public was outraged, the press sarcastically nicknamed the company owner **'Pharma Bro'**, but there was nothing to be done, as the company was legally within its rights.[125]

Italy's **Leadiant** was already selling *CDCA* as a treatment for gallstones, when it was discovered that it was also effective against the metabolic disease CTX. Now CTX is very rare, and that allowed the manufacturer to abuse a European regulation in 2017 by designating the drug as an 'orphan drug', i.e. a drug targeting rare diseases that has limited sales. To encourage the development of new orphan drugs, Europe grants longer market exclusivity to pharmaceutical companies for this purpose. Although *CDCA* was not new at all, **Leadiant** was now able to sell the drug for many years. Promptly the price was raised from €30,000 to €170,000 per patient. For gallstone owners, too.[126]

The German pharmaceuticals company **Medice** registered the drug *dexamphetamines* in 2016, although it had been on the market for years. It has since been sold under the brand name *amfexa* and costs three times as much as before.[127]

Genzyme, part of drug maker **Sanofi** since 2011, withdrew its patent-expired leukaemia drug *MabCampath* from the market when it was discovered that it also worked against MS. In 2014, the drug came back on sale, now under the brand name *lemtrada*, which was 36 times more expensive than before.[128]

The company **Pharmasset** developed an anti-hepatitis C drug. The patent was bought by **Gilead** for almost €10 billion. It was marketed in the United States in December 2013 under the name *sovaldi* for $1000 per pill, which is $84,000 per course. In Egypt, a course of treatment costs about one hundredth of that, but €800 per course is still far too much for most Egyptians (15% of the 100 million Egyptians suffer from hep-C). Already in the first year of sales, 2014, the producer made over €9 billion in profits from the trade in *sovaldis*.[129]

And here is **GlaxoSmithKline** (**GSK**) again. This time with the antidepressant *paroxetine*; brand name: *seroxat*. The manufacturer had known since 2001 that the drug did not work for severely depressed young people. This was established in an investigation called 'Study 329'. Much worse was the fact that the drug can incite suicide. **GSK** knew that too: during 'Study 329', in a test group of 90 adolescents, 11 of them (that is one in eight) made a suicide attempt (compared to just 1 attempt in the control group). The unfavorable conclusions were not published and sales went ahead as usual. Already in the first year, *seroxat* was prescribed for two million American children and adolescents.

The suicide risk only became publicly known in 2015, when 'Study 329' was analyzed again by independent researchers (Noury et al., 2015). *Seroxat* is still on the market; it is one of the best-selling antidepressants in the world. Of the nearly 1 million Dutch people who use antidepressants, over 60% take this drug.[130] Do you too?

If these sickening examples are not enough for you, you can do your own investigation. Take the cancer drugs *AZT* and *erwinase*, for example, or the psoriasis drug *dimethylfumarate* (brand name *Tecfidera*). If you do, take a look at the methods the pharmaceutical industry likes to use. Like *evergreening*, where an expiring patent is revived by a small change that is announced as a great innovation, for example **Roche's** cancer drug *herceptin*, which could suddenly be administered by injection. Or *salami slicing*: if a drug works for several ailments, you register the various applications one by one (in 'slices') so that you can stretch the patent endlessly. And *price gouging*, where the price of a medicine is not calculated in real terms on the basis of the development and production costs – as the manufacturers consistently maintain – but is maximized on the basis of scarcity and patient demand. This leads very regularly to price increases: *Slow K* from **Essential Pharmaceuticals** suddenly became seven times more expensive in 2017, *glivec* from **Novartis** twenty times. In 2019, *lutetium octreotate* from **AAA**, which was bought by **Novartis**, became five times as expensive.[131]

These and hundreds of other examples together show that the pharmaceutical sector, which, together with medical and social institutions, should be the providers of health and good care, is not focused on that at all. It is a branch of industry that, just like all other commercial sectors, is simply focused on profit. The fact that the means to achieve this is the production of *medicines* is secondary: it could just as easily have been *sausages*. Or *bicycle tires*. The health of people is irrelevant to this industry.

Other industries

The above series of examples is a serious indictment of the drug industry. But not only against that one industry.

I have kept stories about quite a few other industries out of this book, because otherwise you could drown in the large amount of facts and backgrounds. But I made them available to you anyway, in the form of five sizeable appendices. So if you like, I invite you to download them.

In Appendix 7, I tell about the crimes of the *tobacco industry*, which has used pseudoscience extensively for many years – and continues to do so. You may recall that in Chapter 6 (page 60) I wrote why this is something entirely different from protoscience? In the case of the tobacco industry, there is absolutely manipulation and deception going on, resulting in millions of deaths.

In that same appendix, after tobacco, the *chemical industry*, the *food industry*, the *automotive industry*, the *lobbying industry*, and the *fossil energy industry* also come into play.

Then, in Appendix 8, I provide further information on the systematic transfer of the risks and consequences of all kinds of activity to, respectively: distant neighbors; nature; and future generations. If all these externalities were included in the cost price of products and services, consumers would pay 'true prices', thus providing our distant neighbors with fairer rewards and less damage. It would also protect the environment and nature, leaving the world cleaner and more beautiful for our grandchildren. But, of course, that pass-through would also annoyingly increase costs for business owners and reduce their profits.

In Appendix 9, I list a number of 'true prices': of meat, air travel, energy, milk, chocolate, and cotton, among others.

And finally, there are: Appendix 10, in which I present some shocking facts and calculations about the *nuclear power industry*; and Appendix 11, in which I do the same for the *financial sector*.

Together, the cases and facts from all these business sectors show that this is not a matter of isolated events. The sum of it all proves adamantly that there is indeed, as I wrote, a rampant disease that pervades the entire global economy: *profit-seeking is a pandemic*.

Fundamental rights

Are you familiar with the American company **SpazeGaze**? It has applied for and obtained a patent from the United States Patent and Trademark Office for the admiration of *astirixia©* in the sky. Anyone in the United States who looks up on a clear night and discerns *stars* will have to pay the young but fast-growing startup US$200 a time. A special rate applies to the so-called *planitoria©*; for example, a view of the red planet *Mars*, now known by its brand name *Aryx©*, will henceforth cost $15,000. The management of the promising company intends to enforce strictly, including the use of drones to monitor compliance by the American people. Violations will be fined.

Patents in Europe, India and Egypt have been filed.[132]

You may understand: **SpazeGaze** does not really exist, the story is fantasy. Stargazing is allowed for everyone; it is a right of all people. But health is also a right of all people.* And so logically the free – or at least accessible to everyone – and safe availability of the means to promote or restore that health should also be a fundamental right of all people, just like admiring starlight. Medicine should have the same status as starlight and the air we breathe; that is, it should be truly available to all. You can compare it to the resources in the soil of all countries: they too should be shared property of everyone, as I argued in Chapter 5 (page 55).

Of course, there is an important difference. Medicines, most of them at least, were invented or discovered through human activity by often large corporations, whereas the air, starlight, and even soil resources were always there. Now if the pharmaceutical industry had shown that they do indeed make their products available to everyone in a realistic, affordable and medically sound way, everything would be fine. There would then be nothing wrong with profit, it would just be a tool for the benefit of business continuity and research. But in the current situation, in this industry 'profit' is an end and not a means: it is '*grab as grab can*'. The result: the unnecessary death or reduced quality of life of millions of people every year.

I'm not spinning a yarn. A careful financial-economic study concluded in 2020,[133] that

> "… large pharmaceutical companies are generating profits for shareholders on a scale that is economically fragile and socially unaffordable. Their payouts to shareholders crowd out essential investments and drive up the price of drugs at a cost that society can ill afford."

With the industry unable to make human health its goal – and profit only as a necessary means to that end – there is only one acceptable alternative. As noted earlier, it is not yet quite possible to already determine what an intrinsically sustainable

* Universal Declaration of Human Rights, Art. 25 (1): "Everyone has the right to a standard of living adequate for the health and well-being of himself and of his family, including food, clothing, housing and medical care and necessary social services…" (UN, 1948).

global economy might look like, until there is solid science to enable us to do so. But one obvious precondition of such a system can already be identified.

> All vaccines, medicines and healing methods should be the shared property of all humanity. Not just in the next century: now.

It is not difficult, even in the current unsustainable economic system, to devise and implement an appropriate structure for this purpose. Others have done so before.[134]

To begin with, all existing patents on medicines should be abolished, perhaps through a transition period. Whether the drug companies should receive financial compensation for this remains to be seen, after all the misdeeds committed by the sector.

New drugs may then still be developed by the existing pharmaceutical companies, which after all have the expertise for them. But payment for this development no longer comes from the sale of screamingly expensive drugs but from public funds, from international organizations such as the World Health Organization (WHO), and from individual governments. The paying organizations and countries identify drug needs, issue international tenders for their development, award the development project to one or more winning companies, and pay them the pre-agreed development costs – and nothing more. Thereafter, the developed and tested drugs are immediately free to produce by any manufacturer or pharmacist, exactly like today's generics whose patents have expired. In short: it is not that difficult.

Undoubtedly the pharmaceutical companies, if they were to hear a proposal such as this, would try to prove by all possible means that it cannot be done. They will be supported by their infantry: pseudoscientists who are paid to generate the desired results. But just as with stubborn children, it is a matter of being firm and persevering.

Profit: a systemic error

In all the examples in this chapter and in the appendices mentioned above, it is repeatedly apparent that the pursuit of profit comes at the expense of individual, social and ecological well-being. There is much more to it than a few isolated incidents. I've called it a disease, *profit-seeking*, and even talked about a pandemic.

It is easy to blame a relatively small number of people, the rich and powerful 'winners' who exhibit 'perverse' behavior. This is convenient, because it can give you the pleasant idea that you would do much better yourself if you were in such a powerful position.

But is that really the case? Here's a modest thought experiment. Find all the perversely acting executives of companies. (Set your own benchmarks to determine who they are.) Do the same with the commercial banks, the stock exchanges, the investment companies, the hedge funds, the international corporations, and, if you will, the governments.

Replace all these 'perverts' with random people you find on the street. (Conveniently ignore that most of them have insufficient knowledge and experience for their new tasks.) Also add yourself to the new crew. *Question: would the financial-economic and commercial world be less perverse in the years that followed?*

The answer is: no, of course not. Perhaps not every fresh replacement would turn out to be a perverse one, but many undoubtedly would. And you: can you guarantee with absolute certainty that you would not succumb to the temptations of an income of millions or even billions, or to the power, honor and luxury that come with it? Maybe. But that doesn't matter much, because there will certainly be many who are less conscientious than you. And it is precisely these people who will work their way up (with their elbows, if need be). After a while, corporations, banks, and governments will be acting just as unscrupulously profit-driven as they do now.

According to many classical schools of economics, humans are *homo economicus*, who – as was shown in Chapter 7 (page 76) – is always and everywhere one hundred percent selfish. If that is true, there lies the explanation for the apparent unscrupulousness of companies and institutions. That would be a bad thing, because then we don't have to hold out much hope that things will ever get better.

Fortunately, this is anything but the case. *Homo sapiens* is a complex being with many, often contradictory, characteristics. Humans are *biological, rational, irrational, emotional, greedy, generous, loving, foolish, wise, playful, poetic, beautiful beings.*

This is reflected in the organizations built by humans, for there are plenty of companies, banks, and countries that, on average, do a pretty good job of operating decently. By no means does every company fit into the series of perverse examples in this chapter.

No, the real problem is not in the character of people. Or that of corporations. But in the fact that the economic, legal and political system has been shaped – thanks to spontaneous and uncoordinated proliferation over the past centuries – in such a way that people's perverse qualities are given opportunities, indeed encouraged. The globally dominant system evokes and reinforces unconscionable profit-seeking.

This is an extremely important conclusion, because admittedly you cannot fundamentally change people; or if you can, at least that is not the ambition of the plan of this book. But systems we can change. For we created them ourselves.

Recommended further reading (for details, consult the online Bibliography)

Book: Jackson, Tim (2009): *Prosperity without Growth: Economics for a Finite Planet. Earthscan, London.*

Book: Daly, Herman E. (2014): *From Uneconomic Growth to a Steady-State Economy.*

Online publication: Fernandez, Rodrigo & Tobias J. Klinge (2020): *Private Gains We Can Ill Afford. The Financialisation of Big Pharma.*

Journal article: Baker, Dean (2004): *Are Patents Necessary? A Free Market Solution for Prescription Drug Crises.*

9

A BITTER SUCCESS

Homo Solitarius, the Lonesome Human

The previous chapter began by defining scientific success. Next, most forms of such success were examined.

One element of success remains: a contribution to the *insight into who we are*. That is the subject of this chapter.

The question of who we are can be approached from many angles, and thus answered in many ways. For example, you can look directly at what most economic schools claim. From that angle, the answer is well known.

> *We are: homo economicus. Units of account in economic models.*

But fair is fair: many economic theorists are aware that this is far too simple an answer. Better, richer answers are therefore possible. Over the course of the chapter, a number of them will come along. Several of them were not developed from economic theory. Some are thousands of years old, coming from ancient religious traditions or impressive philosophical currents.

But they have subsequently been embraced, embedded, and elaborated within the frameworks of economic models, and are widely applied in our time in the dominant capitalist economy.

Some of these responses are usually not formulated explicitly: they seem rather unspokenly present, like family members about whom you'd rather not talk because they're there but you're a bit ashamed of them. I am going to name them.

The answer to the question of who we are, depends partly on what you mean by 'we'. It could include, for example, all living things. Or just humans. *All* humans, including our ancestors and our progeny. Or just the people who are alive today. Only western people. Only the people who matter. Only the rich and powerful,

DOI: 10.4324/9781003306269-10

because they set the rules and control their implementation. Only the obedient, the good boys & girls. Only the winners.

The smaller you define the group called 'we', the more you exclude. The result is *alienation*, that is: distanciation from who you are. This applies not only to the ostracized but also to those who do belong, because they lose their roots and their kin. This is sad, because: the more you reject, the lonelier you end up.

9.1 Alienation from nature

Oh, we know how to explain it perfectly to ourselves: the Planet Earth, with everything on and in it, belongs to us. It's our heritage, our property, our stewardship, our enjoyment, our ownership, or whatever words you want to use for it. I will use *ownership* as a general term, and if you find that legally incorrect, you can flexibly replace it yourself with one or more of the above or with your own choice.

This word 'ownership', along with all those related words, is a fundamental concept of the economy. In Chapter 5, I showed how it can lead to strange phenomena – such as war or robbery – related to the distribution of resources and the exploitation rights of human genes. The restasis story in Chapter 8 is also about ownership and exploitation law. If you talk about economics, then the ownership principle is perhaps the most fundamental concept of all, because many other concepts are derived from it, such as *buying* and *selling*, and therefore *trade*, *money*, *banks*, *capital*, *pay*, *shares*, *derivatives*, *profit*, *tax*, *donate*, *borrow*, *rent*, *lease*, *theft*, *conquest*, *wealth*, *poverty*, *slavery*, all kinds of *fraud* and much more, in short: perhaps the entire economy. Just check: try to define each of these words without (possibly stealthily) using ownership or one of its (rough) synonyms. If a society does not know the concept of 'ownership' or defines it differently than we do, their economy will certainly be organized radically differently.

This is an extremely important observation, because 'ownership' is an impetus word: see Appendix 3 for extensive evidence. So, it could well be that all the words derived from it are, too.

Humans and other animals

Are we thus, possession-based humans, so much worse than other animal species? Well, no. The notion or emotion of 'mine!' is not only found in humans: it exists just as well in other species. Look at a dog that has a bone: have you ever tried to take such a thing from a pit-bull that shows its teeth and growls? All right, dogs are bred by humans. But you can also think of a tiger in the wild: would you dare to take a prey from it? Tigers defended their possessions, long before humans existed.[135]

Areas such as forests, steppes and swamps are also taken into ownership. That happened at least a hundred million years ago, at a time when the mouse-like ancestors of humans still lived modestly in a world dominated by dinosaurs. Ants mark the area of their nest, and they have existed for at least a hundred million years without many changes: they are a long-running success of evolution. The antish territory is defended tooth and nail against peers from other nests, or conquered from them with brutal violence, which shows that war has not been invented by people either.

It is not only ants that go to war. Our closest relatives, the chimpanzees, are also skilled warriors: groups of chimpanzees roam when their habitat has become too small. They fight with the males of neighboring groups, kill some and take their land. They didn't learn this behavior from humans; they initiated it themselves. Jane Goodall, for example, reported on the horrific four-year 'Gombe Chimpanzee War' in Tanzania, which occurred between 1974 and 1978.[136] John Mitani even wrote about a ten-year war between groups of chimpanzees in Uganda, which killed at least eighteen.[137]

But none of those other species are as good at owning as humans are.

We are: the owners of the planet and everything on it. Superior, barely related to the others.

The rightfulness of our ownership

Thus, ownership and taking away existed long before we humans came on the scene. But we have added something new. To begin with, that is the *word* 'ownership', as an expression of the conscious awareness that we are lord and master of things such as: goods, territories, ideas, animals, and even other humans.

A second human innovation is that we try to explain to ourselves why it is 'correct', 'good', 'just', or at least 'rightful' that we own such things. Religious leaders explain to us that it is a God-given right in the form of property or stewardship.[138] Philosophers wrote books to prove our property rights on ethical or moral grounds. Important contributions to the debate have been made by Plato, Aristotle, Thomas Hobbes, David Hume and Karl Marx, among others.[139] They wrote about private ownership, where individual people are the owners of something, as well as shared ownership, where a larger group, for example a country, jointly owns. (For countries, this is not called 'ownership' but 'territorial sovereignty'.)

An interesting position has been taken by John Locke. Following Roman law, he wrote[140]:

> "Though the earth, and all inferior creatures, be common to all men, yet every man has a property in his own person: this no body has any right to but himself.
>
> The labour of his body, and the work of his hands, we may say, are properly his. Whatsoever then he removes out of the state that nature hath provided, and left it in, he hath mixed his labour with, and joined to it something that is his own, and thereby makes it his property."

So, whoever mixes a piece of land with his sweat to work it, automatically becomes the owner of it, according to Locke. Provided that you are the first to do that. And that you are a member of the species *homo sapiens*; otherwise, it doesn't count. Animals that are not humans cannot be owners.

We are: the rightful owners. Because we should be: we proved it ourselves.

The strange thing is that our ancestors themselves were not *homo sapiens*, long ago. Once there was a time when there were no 'wise people' on the planet. But that changed when some primates started walking upright and got bigger and bigger brains. Sometime around then, our 'rightful' ownership must have started. Or was it only later, when the first permanent residences were furnished?

Humankind is at least about 300,000 years old. The oldest archaeological remains of young humanity found so far have been found in Morocco: they are estimated to be 315,000 years old,[141] and they are the first to unmistakably display the characteristics of modern man.

Sometime before, other animal species had separated from 'our' branch of the evolutionary family tree – or we from theirs, because it depends on just how you want to see it. Fourteen million years ago, orangutans and we each went our own way; eight million years ago, the gorillas. Two million years after the departure of the gorillas, there was a branching from which the current chimpanzees and bonobos emerged. Our own branch, that is to say the branch living in Africa from which modern man came, split many times afterwards. Most of the species that emerged from it became extinct long ago: it was a coming and going of great apes and hominids.

Less than a million years ago, when 'we' were clearly quite human-like, our branch split again. This is when, somewhere between 800,000 and 400,000 years ago,[142] the Neanderthal people emerged, left Africa and settled in Europe and Asia.

'We' continued our own evolutionary path and, as said, showed properties characteristic of modern humans, *homo sapiens*, at least 315,000 years ago. Then, just like *homo neanderthalensis*, we swarmed from Africa to Europe and Asia. The oldest traces that have been found so far are in present-day Israel and date from around 180,000 years ago.[143] Somewhere between 120,000 and 80,000 years ago we also entered the area that is now called China, after which Australia was reached, some 45,000 years ago at the latest.

It was probably around 25,000 years ago, during a fierce Ice Age, that America was first inhabited by humans. They came from Asia and arrived in Alaska via the temporarily dry Bering Strait: the sea level was lowered by the massive land ice. That entry is not entirely certain: there are also alternative theories about the first Americans. In Alaska, due to the presence of harsh ice plains, people were unable to continue for thousands of years. But from 17,000 or 16,000 years ago they began a long expansion toward the extreme point of South America, where they arrived around 13,000 years ago.[144]

The last isolated pieces of land followed not so long ago. The Melanesian Islands received human habitation from perhaps 5000 years before now. Hawaii was reached via Polynesia around 400 AD, after which Easter Island probably followed around 800 or even around 1200.[145]

The last step was Antarctica, which was almost discovered by James Cook in 1773, but really only in 1820 by several expeditions. The first person to enter the continent was probably Captain John Davis in 1821. It was only as recently as 1978 that a human being was born in Antarctica for the first time.

And now the whole world is ours.

> *We are: the almighty owners. For we have taken possession of everything; no one could resist us.*

The myth of the noble savage

Thus we, humans, went into the wide world. From Africa to Asia, Europe, the Americas and the rest. We used nature that we found as we liked, without paying attention to the consequences. Of course, there is no blame, because in those prehistoric times there was no way to foresee the unfavorable consequences. Unfortunately, everywhere we went, we caused destruction. Modern storytellers such as Jared Diamond and Yuval Noah Harari have described what happened, every time man first entered a new area.[146] *Their bitter conclusion is: Everywhere where human appears, a regional extinction wave occurs immediately.*

As an example, there were once many large animal species in America. But as the people – the ancestors of the Indians – first traveled through North America from north to south, the North American horse died out, as did the giant armadillo, the mastodon, the mammoth, the cheetah, the lion, the giant sloth, the camel and many other large animal species.* It was not just about mammals, because the giant condor died out, just like the saber-toothed salmon (yes, really) and many, many other species. When South America subsequently received human habitation, the South American horse died out there, as did the largest llama, a hoofed animal, an elephant-like one, a bear and the saber-toothed tiger.† The same things happened when people entered Australia and Madagascar. It was always the largest animals that died out the fastest, on the one hand because they were the most interesting prey for hunting, and on the other because they had the slowest reproduction rate. The animals were not used to humans and were not afraid of them; by the time

* Equus scotti, glyptodon, mammut, mammuthus, miracinonyx, panthera leo atrox, megalonyx and camelops, respectively. The condor was aiolornis incredibilis, and the salmon was posthumously given the name oncorhynchus rastrosus.

† Consecutively hippidion, macrauchenia, toxodon, stegomastodon, arctotherium and smilodon.

because there are plenty of human organizations that can represent them. This is nothing new: the same happens when the rights of children or the mentally handicapped need to be defended.

And nature as a whole: could it be recognized as a legal subject? In the 1960s, James Lovelock proposed that the nature of the entire planet be regarded as a complex, self-regulating system, which to some extent can be understood as a person. Lovelock gave this person the name 'Gaia', derived from the Greek primordial goddess who represents Mother Earth.[160] Lovelock's ideas have been firmly criticized for a lack of scientific or empirical evidence, but they have nonetheless proved influential to how we think about the environment and nature, and our relationship to them.[161]

Whether or not nature as a whole can be seen scientifically or legally as a person, it will be interesting to explore what the consequences would be of conferring to it the status of legal subject, and thus of certain rights. Perhaps: the right to life, or even to freedom from torture and cruel treatment? The right to property? The right to be represented in court? If so, to what do these rights apply: the jungles of Brazil and Sumatra? The North Sea? Animals in the wild? The tiger, as a species? Laboratory animals, pets, livestock?

What would be the impact on an economic system? Massive, I suspect. World-shattering. Ragingly interesting. Perhaps extremely important.

We may be: members of a large, colorful family.

9.2 Alienation from each other

For the time being, humanity is not bothered by such competitors. The other five human species are safely extinct; the great apes almost. Nature is our property, so that is all very pleasantly arranged.

A small space of uncertainty remains. Robots and computers are not yet intelligent and creative enough to demand their share of the pudding.[162] Intelligent aliens have not yet shown up, and the neanderthal has not yet risen again from the grave through genetic engineering, though that is being considered.[163] But wait… bioengineers are practicing building new life forms. In recent years, *chimeras* are successfully being constructed: genetic crossbreeding, including between humans and animals. The technique is still in its infancy, but is already leading to ethical discussions that "… are dominated by concerns that animals with chimeric brains might develop human-like cognitive abilities."[164] If such chimeras are created, will they have human rights?[165] *Who are 'we' from then on?*

Such questions are essential, for the inequality in power and possession *between homo sapiens* and the others is huge. But the inequality *within homo sapiens* is perhaps even bigger, for the economic system is extraordinarily successful in creating wealth – but not equally for everyone.

Inequality between countries

Have a look at the African country of *Malawi*, the poorest country in the world. Then look at *Qatar*, the Arab mini-state that is the richest of all countries. In terms of area, Malawi is about ten times larger, in terms of population about seven times: 20 versus 3 million people. But Qatar's GDP is higher: 27 times that of Malawi. Per capita GDP in Qatar is more than US$66,000, compared to $342 in Malawi.

The immense difference comes from fossil fuels. Qatar has rich oil and gas fields, producing one and a half million barrels of oil per day and also 450 million cubic meters of natural gas. Malawi has nothing. And so, Qataris enjoy an average of 78 years on an annual income that is two hundred times that of a Malawian, who fortunately only has to live in poverty for an average of 64 years.

Try to imagine what that looks like. Think of a gymnasium with two groups of people, one on each side. One 'group' consists of one person, a man (*surely a man*) with a keffiyeh, a cloth wrapped around the head. The other consists of two hundred people, all dark-skinned natives of Malawi. Can you see it? Both groups have the same combined income. Now, just for fun, subtract the necessary living expenses for both groups from that income, and what remains is the disposable income. There is every chance that that amount for the Qatari 'group' is now *two hundred thousand* times that of the Malawian.

Qatar's wealth enables it to make gigantic expenditures. This is done in part through investments in the rest of the world, with the largest investments being made in real estate and western companies in London, New York and Washington, DC. In 2005, the *Qatar Investment Authority* (QIA) was established, a state-owned enterprise that invested US$335 billion abroad in just twelve years.[166] For mini-state Qatar, that is only one-sixth of the GDP of those twelve years; for much larger Malawi, it would have taken the total GDP of fifty years. "Qataris own more of London than the Queen", the *Telegraph* headlined in 2017.

In addition, Qatar is fond of prestige objects and projects, with the notable example of hosting the 2022 World Cup in a country that is completely unsuited to it in terms of climate and soccer tradition. In the construction of the stadiums and other facilities, foreign workers from India and Nepal, among others, lived like slaves; at least 6,500 of them have already lost their lives, *The Guardian* reported in February 2021.[167] Qatar shares its love of heavy spending with the similar *United Arab Emirates* (UAE), which apparently doesn't know what to do with its money and so puts a large portion of its petrodollars to work in the game 'Who's got the Largest?' In 2008 they won: with the Burj Khalifa in capital Dubai, 830 meters high, pictured in Figure 9.1. They are warmly congratulated.

We are: the winners. We won fairly from the poor nations. Through our trade, and well, also through war and colonialism. Back in the day.

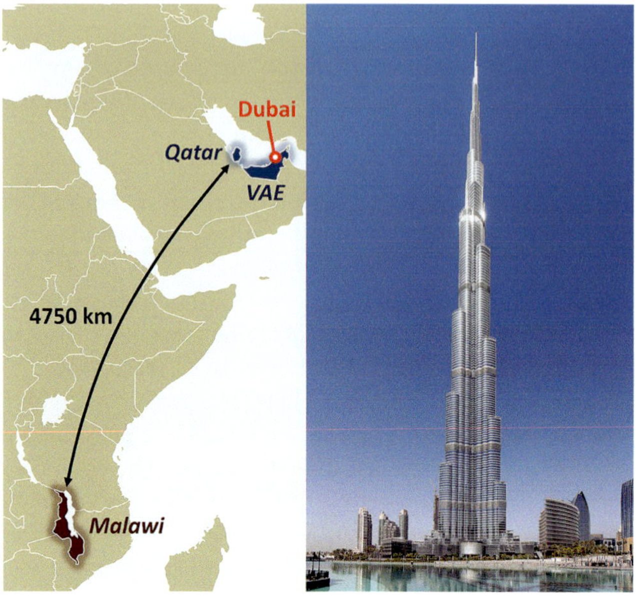

FIGURE 9.1 Left: Malawi, Qatar and the UAE with capital Dubai. Right: The winner is: Burj Khalifa in Dubai.

Inequality between individuals

The highly unequal distribution of resources between countries can lead to war, as the example of the Republic of Groningen (actually Biafra) showed. But the inequality between *individual* people is even more absurd. One Qatari outweighs two hundred Malawians? Ha, that's peanuts. How about one wealthy person to *one hundred and forty-six thousand* less wealthy people? Oxfam reported it. In 2019, the 26 richest people together owned $1.4 trillion (UK: billion), and that's as much as the summed assets of 3.8 billion others (3.8 billion divided by 26 ≈ 146,154). That's half of *all* the people on the planet.

That even surpassed Oxfam's 2016 report, in which it took the 62 richest people to be as wealthy as the poorest 3.5 billion people.[168] In just three years, inequality had become even more extreme. No wonder, because: the billionaires together become US$ 2.5 billion richer *every day*.[169] But why, actually?

The rich are getting richer, and that is perfectly logical. Because wealth and power have positive feedback, a self-reinforcing effect, thanks to various mechanisms. A number of these are shown in Figure 9.2. But poverty also exhibits positive feedback: those who are poor have a higher-than-average chance of remaining poor or even becoming poorer – and passing that trait on to their children.

And when the coronavirus crisis broke out in the early 2020s: what happened next? Again, Oxfam provided the answer.[171] One of the conclusions in 2021 was that although the richest thousand billionaires lost 30% of their wealth in the

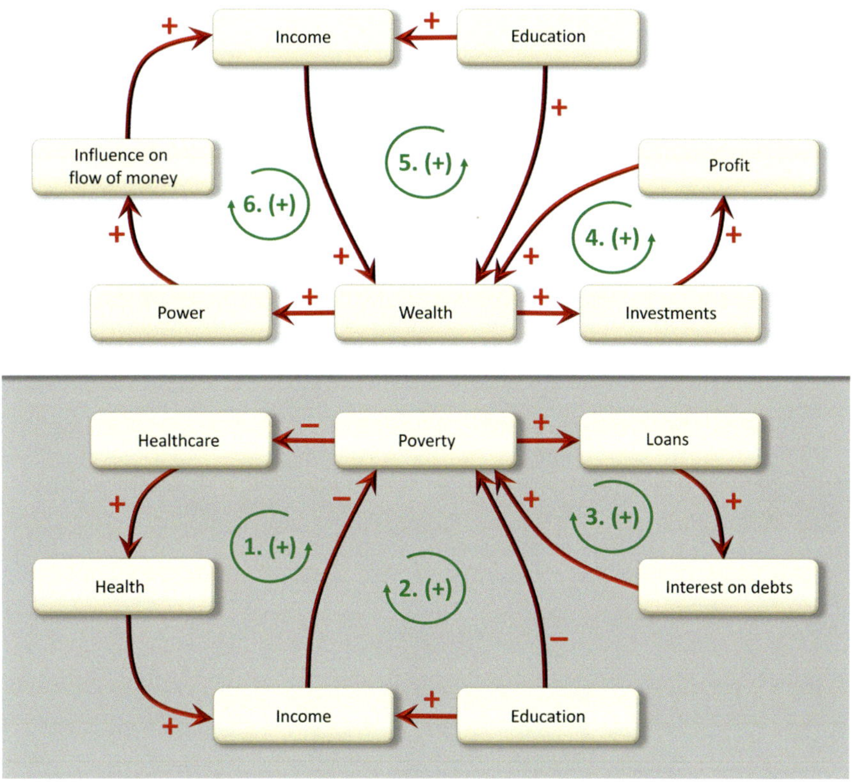

FIGURE 9.2 Rich makes richer, poor makes poorer.[170] Explanation: the '+' from healthcare to health means that the two are 'positively correlated'. If one goes up, so does the other; but if one goes down, so does the other.

first few months of the Covid-19 pandemic, they had already recovered fully by November: they were as rich as before. The wealth *growth* of the richest ten persons during the first year of the crisis was sufficient to pay for the vaccinations of the entire world's population. (Need I say it? It didn't happen.) On the other hand, the poorest are expected by the World Bank,[172] quoted by Oxfam, to take at least ten years to climb back even to their old level of poverty.

What do you think: does it take even harder evidence to demonstrate that there is something completely wrong with our conception of 'ownership'? *Rightful* ownership?

Money attracts money. Money clumps. I have depicted this in an illustrative way, as you can see in Figure 9.3. In it, it is assumed that initially there is a perfectly even distribution of money (Figure 9.1), after which it clumps together more and more (Figures 9.2 through 9.4), causing the super-rich and powerful to emerge.

This continues until the inequality becomes unbearable and a tearing revolution breaks out which causes everything to fall apart: think of the French or Russian

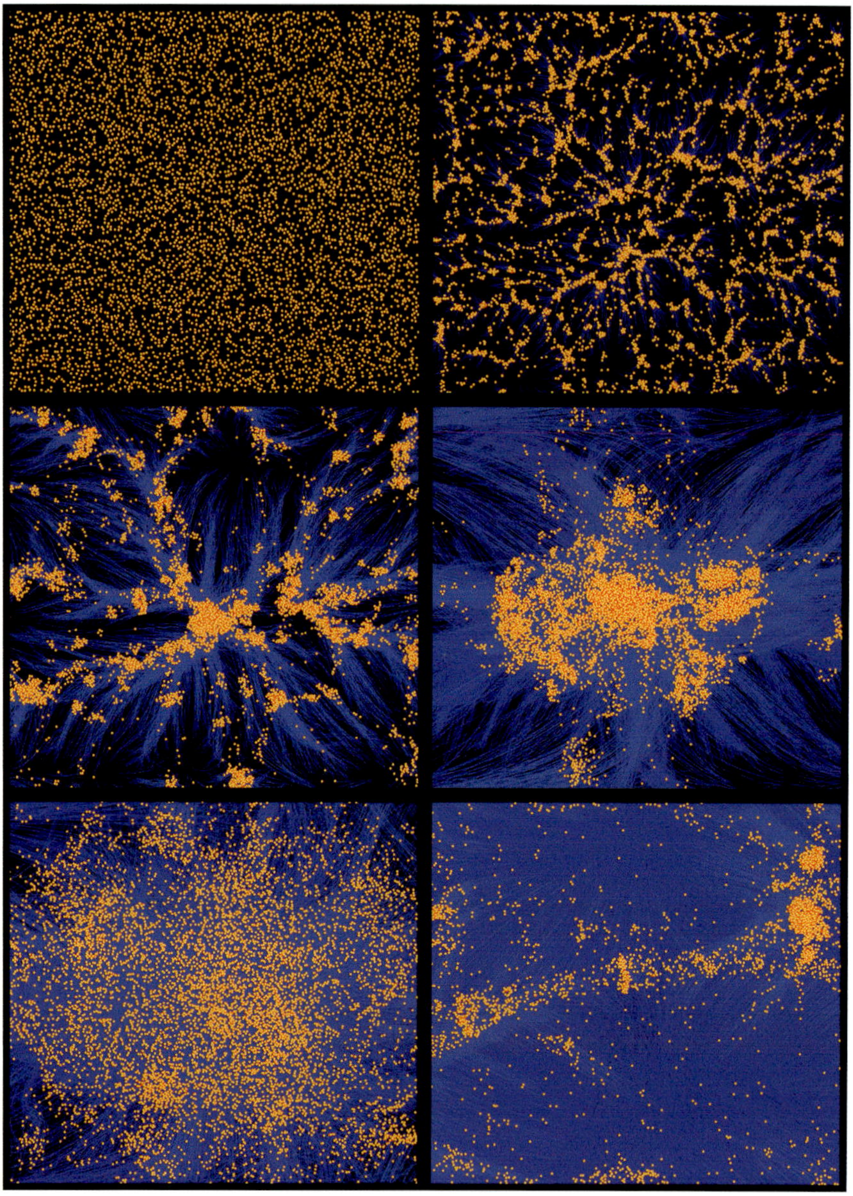

FIGURE 9.3 Money attracts money, as symbolically shown in a film clip you can watch on YouTube, made with a computer program you can download.

Revolutions. After which the money begins to clump together again but in other places.

In this 'comic' you can choose to see the development of: the distribution of money; the distribution of power; the clustering (mergers) of companies, leading to supreme multinationals; the emergence of superpowers such as the Roman Empire or the United States; but also, for example, urbanization. Naturally, if you wish, the video clip also visualizes physical processes such as condensation, crystal formation, or the creation of galaxies as a result of the contracting gravity that caused the matter in the universe to clump together shortly after the Big Bang.

Rich makes richer; poor makes poorer. Money attracts money, and so inequality increases structurally. That this principle is not fiction is shown by thorough research, in both the shorter and longer term. The latter was confirmed by an international study led by Tim Kohler, which shows that inequality between rich and poor has been on the rise for 10,000 years.[173] That inequality is expressed in the *GINI coefficient*, which theoretically can vary between 0 (complete equality: everyone is equally rich) and 1 (maximum inequality: everything is owned by one person). Based on archaeological findings, Kohler and his colleagues concluded that the GINI in the Neolithic was approximately 0.17.

As a result of the first small-scale agriculture, which led to permanent settlements and thus to more property, inequality rose to 0.25 and then, as livestock herds grew, to 0.35. In the Middle Ages, the power of nobility and clergy increased and the class of landless peasants grew, with inequality growing to 0.56. In modern times, according to the financial institution Allianz SE, the GINI coefficient has values like 0.53 (China); 0.64 (Netherlands); 0.73 (Brazil); and the highest value: 0.81 (United States):[174] extremely unequal, higher than ever before.

In the shorter term, of course, inequality does not rise continuously: fluctuations can be seen over time, in terms of inequality both within and between countries. But since about 1980, inequality has been growing at record rates. This is pointed out, among other things, in a book by Frenchman Thomas Piketty that attracted a lot of attention.[175] Based on large amounts of observational data, the author proved that the return on wealth is higher than the return on labor, which is a neat formulation of the principle 'rich makes richer'.

To illustrate this: in 2020, the first year of the coronavirus crisis, the number of billionaires increased by 493 to 2755, according to business magazine *Forbes*. Their accumulated wealth rose even more steeply: from nearly $8 trillion to more than $13 trillion.[176]

We are: the rich and powerful. The economic capital is ours, we earned it ourselves. Using our economic capital. Or inheriting it from our parents.

But what am I complaining about! Poverty is declining, right? Thus, several authors happily reported to us. For example, Hans Rosling in his book *Factfulness*,[177] and Steven Pinker in *Enlightenment Now*.[178] Both books have become famous in 2018, but they completely miss the point. In reality, things are a lot more complicated, as I show in Appendix 13. Poverty is falling – but only in the eyes of those who would like to see it that way. Poverty, also an impetus word?

9.3 Alienation from society

In 2018, a discovery was made in Brazil.

> "**Never before seen Amazon tribe caught on drone video**
>
> An isolated Amazon tribe with no known contact with the outside world has been spotted by a drone flying over the Brazilian jungle, according to the country's National Indian Foundation. Footage released this week shows several people walking through a wide clearing made in a patch of dense jungle in the Javary River valley, near the border with Peru. One of the figures is carrying a spear or pole of some kind, while four or five others stand near what look like thatched structures. None of the people spotted by the drone appear to notice their observer, which is flying high above the trees surrounding them."[179]

This discovery was not the first, and probably not the last of its kind.

> "Brazil's Amazon is home to more uncontacted tribes than anywhere in the world. There are thought to be at least 100 isolated groups in this rainforest. Their decision not to maintain contact with other tribes and outsiders is almost certainly a result of previous disastrous encounters and the ongoing invasion and destruction of their forest home. Very little is known about these peoples. What we do know is that they wish to remain uncontacted: they have shot arrows at outsiders and airplanes, or they simply avoid contact by hiding deep in the forest."[180]

Loss of cultural identity

Now that we've found these people, what do we do? Do we leave them alone? That's unlikely, history shows. Or are we going to inform them of our modern 'civilization'?

That's a damned tricky question. On the one hand: who are we, western people, to attack or perhaps even destroy the hundred or two hundred small cultures that still manage to survive just here and there in their original state? On the other hand: who are we, if we decide for the children of such tribes that they are not going to have access to our western achievements such as writing, mathematics, cell phones and energy drinks, without asking them what they themselves wish and simply leaving them to the views of their parents?

Presumably this crazy ethical dilemma is irrelevant in practice, as it seems very likely that the process of eradicating these ancient cultures is unstoppable – not to mention literally eradicating large portions of these people themselves, as they usually fall prey to infectious and for them deadly diseases such as measles, at the very first contact. The process seems unstoppable, if only because their habitats, especially the jungles of Brazil and Sumatra, are shrinking rapidly.

The disappearance of cultural habitats and the absorption of primitive tribes are not the only causes of cultural loss. Thousands of cultures and languages are currently wearing away, largely losing their cultural identity as they blend with the ever-expanding dominant western culture. Many local cultures will eventually merge into the massive unified culture and become invisible – or have already done so. Just as biodiversity is declining at lightning speed in what biologists call the 'sixth extinction' (see next chapter), so cultural diversity is currently declining at a rate that, I suspect, has never been seen before. There is a veritable 'cultural extinction', as Figure 9.4 illustrates with a number of 'photos' of endangered languages and

Deg, naw, eth, seyth, hwegh, pemp, pajar, teyr, dew, huni, mann.
Cornish, South England: critically endangered, 557 speakers.

Ne'wtisga'q, pesgunateg, ugumuljin, lluigneg, as'gom, na'n, ne'w, si'st, ta'pu, newt.
Mi'kmaq, North America: endangered, 15,000 speakers.

옴뿔루, 시우아, 활루, 삐쭈, 노오, 을리마, 빠아, 똘루, 루아, 디세.
Budonese, Sulawesi: endangered, 79,000 speakers

ᏍᎪᎯ, ᏐᏁᎳ, ᏧᏁᎳ, ᎦᎵᏉᎩ, ᏑᏓᎵ, ᎯᏍᎩ, ᏅᎩ, ᏦᎢ, ᏔᎵ, ᏌᏊ, Ꮭ ᎪᏍᎫ.
Cherokee, North America: endangered, 11,000 speakers.

Manot, sigua, gualu, fiti, gunum, lima, fatfat, tulu, hugua, hacha.
Chamorro, Oceania: vulnerable, 62,500 speakers.

Дæс, фараст, аст, авд, æхсæз, фондз, цыппар, æртæ, дыууæ, иу, нуль.
Ossetic, Caucasus: vulnerable, 500,000 speakers.

Mari, aylla, pura, reqle, kayu, kechu, meli, küla, epu, kiñe, sero.
Mapudungun, South America: critically endangered, 200,000 to 400,000 speakers.

थिम्, हिम्, हिम्, स्यम्, च्यम्, कुम्, पिम्, तिय, दुद, किऔ, लिन्‍ग.
Dura, Nepal: extinct in or shortly after 2008 with the death of Soma Devi Dura.

Ten, nine, eight, seven, six, five, four, three, two, one, zero.
English, Western Europe and America: not endangered, 335 million speakers.

The threat categories of the 15 cited languages is derived from the 'Red List' of languages: *UNESCO Atlas of the World's Languages in Danger* (UNESCO, 2018).

Numbers relate to those who use the language as native speakers.

Ngaal, wagaa, janya, duwa, jugu, marla, daan, guga, bularri, garlugun.
Gumbaynggirr, Australia: critically endangered, 40 speakers.

איינס, צוויי, דרײַ, פֿיר, פֿינף, זעקס, זיבן, אַכט, נײַן, צען.
נול
Yiddish, Europe: endangered, 1.5 million speakers.

ვით, ცხორო, ოვრო, შქით, ანსი, ხუთ, ოთხ, სუმ, ჯურ, არ.
Laz, Turkey: critically endangered, 22,000 speakers.

ʕaʃrˁa, təsˁʕa, ətmanja, səbʕa, sətti, χamsa, arˁbˁəʕa, ətlata, ətnain, waнəd.
Siwi, Egypt: endangered, 15,000 speakers.

Kim, īdõks, kōdõks, seis, kūž, vīž, nēļa, kuolm, kakš, ikš, null.
Livonish, Latvia: extinct on 5 June 2013 with the death of Grizelda Kristina.

ワン，シネペサン，トゥペサン，アラワン，イワン，アシㇰネ，イネ，レ，トゥ，シネ.
Ainu, Japan: critically endangered, 15 speakers.

FIGURE 9.4 Threat levels of some languages.[181] One of these languages is unthreatened.

scripts. It is not just the newly discovered Amazonian Indians who are in danger of being uprooted. All over the world, people are alienating themselves from their identities as a result of the global spread of western culture.

> *We are: the top culture. We have created unparalleled knowledge, technology, wealth, language, means of communication and civilization. We are spreading it all over the Earth.*
>
> *Pleasant side effect: thus, we open new markets.*

Loss of human dignity

Isn't it silly of us that we did not even realize that we had a need for rooms in which we could alternate between bright pink, purple or green lighting? Fortunately, a national newspaper brought us up to speed.

> "*It used to be simple. You controlled the lights in your house with a switch. But that was in the past. Nowadays, even the simplest light bulb contains a mini-computer, and often even wireless internet. Such light bulbs can be dimmed remotely, or can adopt millions of colors. They talk to each other, to the doorbell and the smoke detector. They know when the Sun rises and sets and if it is about to rain. They know if you're home and if they need to be flicked on because there's a nefarious person in the yard. If the resident is having a party, the entire lighting plan flashes brightly along with the music.*"[182]

"New markets", the latest definition of 'we' just mentioned. Creating new markets is part of the game that the great entrepreneurs play to win. That game, of course, includes: creating new needs, like purple lamps. Goal: profit, power, prestige. Game material: us. The *other* 'we'.

> *We are: pawns in the game of the rich and powerful. Of the big banks, of the companies in the Fortune 500, of the billionaires in the Forbes 500.*

Numerous means are employed for that purpose. Open or covert advertising.[183] *Influencers*. Mass media, including Facebook and Twitter. Deepfake and alternative truths. Contempt of science; abuse in the form of pseudoscience. Search engines, such as Google, based on algorithms that manipulate our searches. Trade in big data: *our* data, therefore *our* property, and yet sold by others to others.

Curiously, these violations of our privacy by internet companies are considered disturbing by only a few citizens.[184] Almost immediately after Facebook founder Mark Zuckerberg lied in December 2018 "We've never sold anyone's data," a UK parliamentary inquiry committee found: "Facebook was willing to override its users' privacy settings in order to transfer data to some app developers," for example, Netflix and Spotify, and "to starve some developers of that data, thereby causing them to lose their business. (…) Facebook intentionally and knowingly violated both data privacy and anti-competition laws."[185] The commission described Facebook as "digital gangsters in the online world, considering themselves to be ahead of and beyond the law." Facebook probably takes the crown, but Google, Apple and Samsung, among others, have also paid billions in fines for misconduct.

> *We are: consumers. Manipulable profit machines; tradable data.*

On the other hand, we are employed to manufacture the products that, as a result of our pent-up consumer needs, we want to own. In the process, economic theories are used to study in a quasi-scientific way how we can be utilized in the most efficient way: theories that speak not of people but of human capital, combined if necessary with other concealing words such as reorganization, automation and factors of production.* Workers' rights, fought for over a hundred years through trade unions, are being phased out again; permanent appointments and tenures are being replaced by flexible contracts and the use of pseudo-self-employed workers. Employees accept this, partly because of the threat of outsourcing to low-wage countries without minimum wages and environmental concerns.

> *We are: factors of production. Human capital.*

When a CEO, or a business theory, condescendingly labels employees as 'human capital', it's a bit like what military leaders call 'collateral damage', a famous euphemism for civilian casualties of acts of war.

> *We are, in case of bad luck: collateral damage.*

Both terms help to objectify people, to turn us into *things*. To forget that we have emotions, to deprive us of our human dignity. This makes it easy for the CEO to lump workers in with machines, materials and supplies: all factors of production. That's convenient, especially when you replace your workforce with 'other machines' or make them work under poor conditions: unsafe, unhealthy,

* For example, the famous management method Six Sigma mentions a row of five production factors, the '5 M's': *Machines, Materials, Men, Methods, Mother Nature*. (The living planet is thus also a production factor.) Other M's are also mentioned in addition or in their place, including *Measurements, Management, Maintenance, Mission, Money* (see: Bradley, 2016).

monotonous, insecure, without prospects, poorly paid, many hours and days, child labor. Or even linked to forced accommodation and forced shopping, where the entire salary is compulsorily repaid to the boss, so that the employees are effectively wage slaves with no prospect of improvement. *In 2017, there were 40 million slaves in the world.*[186] Slavery! In the twenty-first century?[187]

Contempt. People who knowingly bear responsibility for such abuses harbor contempt for their fellow human beings. They feel that they are superior to them. They live in *hubris*. They imagine themselves to be gods, for whom the lesser people are irrelevant, except insofar as they are useful as 'pawns' to achieve victories in the economic game of business.

> *We are: gods. On our Olympus, we are untouchable, infinitely elevated above the common people, and even further above nature.*

The rapidly growing inequality and the apparent inviolability of the mega-rich is creating a society in which solidarity is one of the victims. This leads to alienation: people increasingly feel like *'strangers in a strange land'* that they used to think they knew, but now know less and less.

This leads to hatred, both towards the rich and powerful and towards arriving refugees, and subsequently it leads to discrimination, violence and terrorism by both natives and newly arrived refugees or their children.

Then, as fear strikes, the government spies on more and more citizens, making basically all inhabitants of a country potential enemies of the state. This government's appropriate and inappropriate invasion of privacy causes more distrust.

> *We are: risks. Potential enemies.*

Meanwhile, western society is becoming more impersonal and less human. Companies are getting bigger: at supermarket chain Walmart, 270 million customers a week, assisted by 2.3 million employees, generate an annual turnover of US$500 billion.[188] Cities are getting bigger: There are already cities with populations of close to 40 million people. Social networks are getting bigger: Facebook is growing by six new accounts per second and is visited daily by one and a half billion people, each of whom has an average of 338 'friends'.[189] That was only a mere 155 in 2016:[190] are friendships increasing?

The only thing that is not getting bigger is the individual human being: they are still on average smaller than two meters tall. Because everything around them is growing, individual humans are becoming proportionally smaller and smaller: ***humans are shrinking***.

> *We are: shrinking. Drops in the ocean.*

Communication is taking place less and less in direct contact ('Face to Face', F2F) between people. We meet our rapidly growing number of friends via a mini-screen and a virtual keyboard. We do our shopping online, and when I do my household shopping (at a mega-supermarket of a fast-growing chain of stores), I scan my groceries myself, pay efficiently at a payment terminal, don't speak to a single person, and hate myself for it. We report a crime to the police online without an 'arm around your shoulder', we get money from a wall. Everything is automating. Everything is becoming more and more efficient in our 'Lonesome Century'.[191]

Of course: during the coronavirus crisis that began in 2020, we were happy with all these digital contacts. That was only natural, because it was the only thing we had left. But at the same time, many people discovered that this kind of communication is only a limited surrogate of real encounters. Non-verbal communication is seriously hampered by it. You can't hug and pass around beers by computer, which proves:

> *We are still: biological beings, co-driven by genes, hormones and instincts, although we increasingly try to deny this: humans are dehumanizing themselves.*

9.4 Conclusion

Man, what fences and walls we have erected! We have shut out nature. Deprived all animals of their rights. Distanced ourselves from our relatives like the chimpanzees and the gorillas. Showed the poor nations the door. Made the poor individuals into unwanted aliens; or into modern-day slaves, a kind of talking animal who must shut up. Ousted cultures other than the western one. Degraded middle-class people to objects of profit, data sets, factors of production and risks. Promoted ourselves to gods: wealthy and lonesome on our Olympus. Everyone is alienated from everything and everyone.

> *We are: homo solitarius, the Lonesome Human.*

Or, as a contemporary bard sang:

There's so many different worlds
So many different suns
And we have just one world
But we live in different ones

Dire Straits: Brothers in Arms, 1985

In the course of this chapter, I have provided a long line of definitions of 'we'. I began by noting that several of these were not developed from an economic theory. Some are thousands of years old, originating in ancient religious movements or impressive philosophical currents.

But they have subsequently been embraced, embedded, and elaborated within the frameworks of economic models, and are widely used, especially in the dominant capitalist economy. The result is not, as prevailing theories claim, *homo economicus* but *homo solitarius*.

The fact that the economic discipline accepts and strongly reinforces the images of humans that come from religions and philosophical schools, once again emphasizes that economics is far from being a science; rather, it is a mixture of science with religion, philosophy, magical thinking. In short: a protoscience.

The final step in proving that thesis – that economics is a protoscience – has been provided in this chapter. For in answering the question of whether economic theories are successful in contributing to an understanding of who we are, it has been shown that economics provides only unsatisfactory, incorrect, saddening, harmful and even life-threatening answers to that question.

From a valuable, mature science, one might expect something entirely different. I closed the previous chapter with words which I now extend somewhat:

> *We are: biological, rational, irrational, emotional, greedy, generous, loving, foolish, wise, playful, poetic, beautiful beings; living in the midst of our brothers and sisters, animals and plants, in our home: Planet Earth, of which we are a part.*

An overly oppressive economic protoscience is not capable of helping us to make that definition come to life. But wouldn't it be wonderful if we could create a comprehensive omniconomic science that would help us realize this dream about 'we'?

Recommended further reading (for details, consult the online Bibliography)

Book: Diamond, Jared (1997): *Guns, Germs, and Steel: The Fates of Human Societies.*

Book: Harari, Yuval Noah (2014): *Sapiens: A Brief History of Humankind.*

Journal article: Knoppers, Bartha Maria & Henry T. Greely (2019): Biotechnologies nibbling at the legal "human".

Book: Piketty, Thomas (2013): *Le Capital au XXIe siècle.* English (2014): *Capital in the Twenty-First Century.*

10

COLLAPSES THAT MAY BRING OUR FINAL DEMISE

As part of the examination of economics as a protoscience, I have spent the past two chapters looking for the extent to which the discipline of economics has led to successes. In doing so, it became apparent that the success achieved, in the form of an increase in wealth for a select portion of humanity, is accompanied by great harm.

Unfortunately, it is much worse than the previous chapters showed. There is a real risk that the damage will culminate in massive disasters. In catastrophes of global proportions, which threaten the survival of civilization and perhaps of humanity.

These catastrophes are the subject of this chapter. After completing this, I will conclude the first half of the book with a chapter in which I draw the conclusion about the status of the discipline of economics.

Honored reader: this chapter is not cheerful, but regrettably it is necessary. Soon the demolition work of the first half of the book will give way to the constructive work of the second half: toward a new science. That will be more hopeful and a lot more enjoyable. So, hang in there a little longer.

10.1 Growth patterns

Boy, was Malthus vilified. And yet it sounded so logical, what he wrote in 1798 in his book *An Essay on the Principle of Population*. In short, his main thesis was that in the long run something must inevitably go completely wrong. This is because people, given the chance, multiply at an increasing rate, at a fixed percentage per generation: a growth phenomenon called ***exponential*** by mathematicians; the word was already mentioned on page 94 when the actual growth of the world's population was described there.

This would not be so bad, were it not for the fact that, according to the learned preacher Robert Malthus, the production of food cannot grow exponentially but only ***linearly***. You can see both types of growth in chart A of Figure 10.1. And

DOI: 10.4324/9781003306269-11

then it may well be that food production is initially more than enough to feed all people. In fact, food production may initially even increase faster than the number of people. But in the long run, exponential growth will win out over linear growth. *Always*, because that is a mathematical necessity. And so, humanity will one day face a catastrophe. Malthus does not seem to have been a happy man.

The 18th-century cleric was right in that exponential growth does indeed inevitably win out over linear growth. Exponential growth increases per unit time by a fixed multiplier, and that produces a spectacular increase that goes faster and faster: more than almost anyone intuitively expects. Appendix 5 shows something of that. As a result, the exponential curve in Graph A moves rapidly toward infinity and, after a short or long time, beats any linear growth.

Now, a perfectly unhindered growth towards infinity is never possible in reality: such a thing is only possible in theoretical situations, that is, in mathematics. In the real world, sooner or later, there will always be a limitation that will cause growth to slow down. This can happen in three ways.

The first form of decelerating growth is shown in Graph B of Figure 10.1. This is ***logistic growth***. Such growth initially begins in a way that is indistinguishable from exponential. But at some stage, the constraint begins to assert itself. Take as an example a bacterium that likes to live in a pond. Bacteria don't have to deal with complicated reproductive issues, such as finding a mate. So if one bacterium falls into an otherwise bacteria-free pond, it will readily divide and thus multiply, and then there are two. These do the same thing, and thus a happy exponential

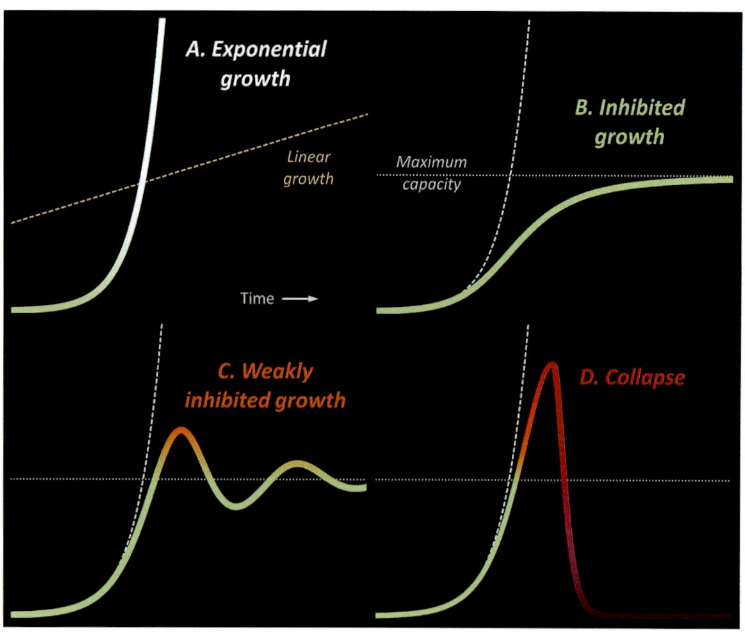

FIGURE 10.1 Various growth patterns.

development occurs. But at some point, near the edges of the pond there will be a lot of bacteria, and that makes it more difficult to find space and food, and thus to prepare for reproduction. The rate of growth decreases and gradually comes to a halt. The population size then reached is apparently the *maximum capacity* of the pond. Logistic growth is a neatly ***inhibited growth***, where the limited size of the physical environment acts as a brake.

Now it may be that for some reason the limiting brake is not very strong. Perhaps, for example, because there is so much momentum in the growth process that the system overshoots, like a heavy truck with bad brakes overshoots when the driver tries to stop at a traffic light: then you see ***weakly inhibited growth***. It may also be because a growing population simply does not notice in time that the limits of the system are near, so growth continues unimpeded for a time. In the metaphor of the truck, it then overshoots its limit because the driver did not start braking in time: this is not weakly inhibited growth but ***(too) late inhibited growth***; the effect is the same. Shooting through beyond maximum capacity is shown in Graph C of Figure 10.1. In this graph; the curve moves up and down several times after the first overshoot. This is not necessarily the case; overshooting growth can also quietly drop to maximum capacity in one go, immediately after the first overshoot.

Graph C is also shown in Figure 10.2, which depicts the computer program *PopSim*. With that program, which you can download from this book's website, you can simulate and examine the various graphs of Figure 10.1.

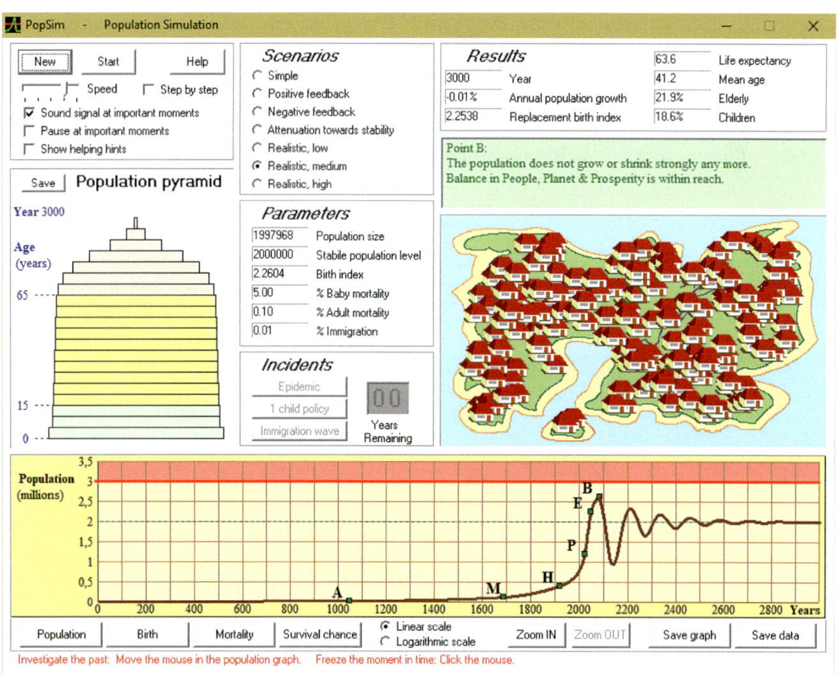

FIGURE 10.2 The computer program PopSim, showing the scenario of Graph C.

In such an overshoot, the system in which the growth takes place may be damaged because boundaries are passed. Some form of overexploitation then occurs. If it is not too severe, the resilience of the system will probably absorb and repair the damage. But if the maximum capacity is grossly exceeded, there is a risk of permanent damage. What can happen then is shown in Graph D: ***collapse***.

Figure 10.1, with its four growth patterns, will play an important role in the coming pages. This makes it worth giving it a name: the figure will be referred to as the *'Four Growth Curves'*. (See Appendix 5 for a description of how the four graphs are mathematically constructed.)

10.2 Theoretics of catastrophe

On page 94, world population growth was shown in Figure 8.1. That graph shows the expectation that growth will slow down over the course of the 21st century. As such, the curve resembles parts of Charts B, C, and D of the *Four Growth Curves*. This begs the question: which of these three graphs best describes reality?

Unfortunately, it can no longer be graph B, because we know that we have already shot through the maximum capacity: the ecological footprint is considerably larger than the biocapacity. So, we find ourselves in Graph C or Graph D. Are we heading for weakly inhibited growth that still bounces back to a healthy equilibrium, or is everything going to collapse? The answer is impossible to calculate, because these kinds of processes are far too complex for that. But a related question can be explored to some extent. If a system collapse is going to occur as outlined in Chart D, how will it happen? how will that work out?

To answer that question, curve D of the *Four Growth Curves* can be seen enlarged in the left half of Figure 10.3. It looks like a neat graph, which gives a far too simple picture of a dramatic and chaotic collapse. No wonder: it shows only one quantity, the population size. The graph would be better drawn as in the right-hand section of this figure, where the various curves represent a variety of system magnitudes, ranging from, say, ocean water quality to the average income of the world's people, or economic productivity. When a system collapses, first one or a few quantities go up

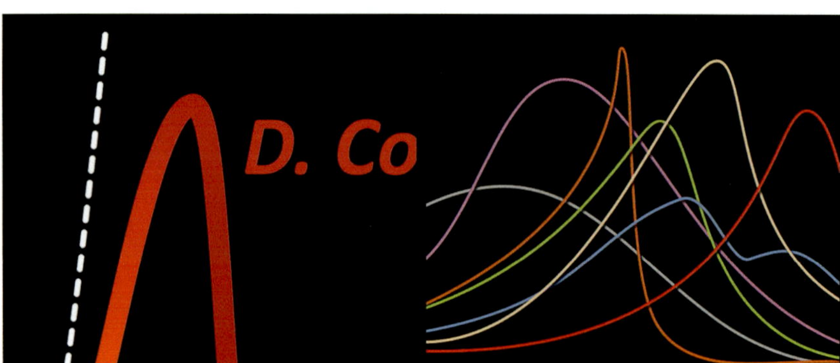

FIGURE 10.3 The transition from too weakly inhibited growth to collapse.

or down dramatically, followed by many others as a result of all kinds of interactions. Bit by bit, slowly or suddenly very quickly, the system crumbles and then collapses.

This right-hand section of Figure 10.3 still looks like a neat picture. But, of course, it says nothing at all about what exactly happens if a large system does indeed collapse. About what people and other animals then experience. Feel. Think, if their world collapses before their eyes.

Scientific researchers can't really help on that score. But they try: Tainter, for example, examined collapse processes and shared that they are defined by some as 'economic disintegration', but by himself rather as "a rapid, significant loss of an established level of sociopolitical complexity."[192] Other authors described collapses in a similarly academic way,[193] which of course is their job, but which doesn't really help one's empathy.

I can't really help you on that score in this book either. But there have been legions of thrillers, science fiction books, feature films and TV series in which attempts have been made to paint a living picture of such disaster scenarios and dystopias. This is done, for example, in the shocking book *The Collapse of Western Civilization: A View from the Future.*[194] There are documentaries of all kinds. Perhaps you can use such sources, along with your own imagination and empathy, to imagine how systems collapse. In any case: chaotic, unpredictable, violent and heartbreaking.

The scenarios of World3

Although you cannot predict the future, especially if it is likely to be chaotic, there is something else you can do: simulate reality as closely as possible in the computer, and then see what might happen. That was first done by Dennis Meadows et al. in 1972, who designed a computer model for it called 'World3'. I mentioned Meadows already in Chapter 8 (page 98), where I said that their pioneering research was vilified by many economists.

In the mathematical model, Meadows included an enormous number of formulas, taken from science and technology fields such as physics and chemistry, geology, biology, environmental science, economics, political science, medicine. He combined this interdisciplinary approach with as much data as were available at the time. He then had the computer calculate what could happen in the next hundred years or so, based on various assumptions ('scenarios') about how humanity and the economy would develop and how people would react to impending disasters.

What emerged? Some scenarios turned out to be dramatic and disastrous over the course of the 21st century, comparable to curve D; others ended happily and reached a pleasant equilibrium, as in curve C. I will show an example of both. 'Scenario 2' (Figure 10.4) is a disaster scenario, arising from the 'Business as Usual' ('BAU') assumption: note, for example, the curve of the population. 'Scenario 9' (Figure 10.5), arising from firm global policy changes, proceeds more amiably, showing weakly but still timely inhibited growth.

The 'Report to the Club of Rome',[195] in which the Meadows team discussed the results of their simulations, had an explosive impact. Pseudoscientists did not let

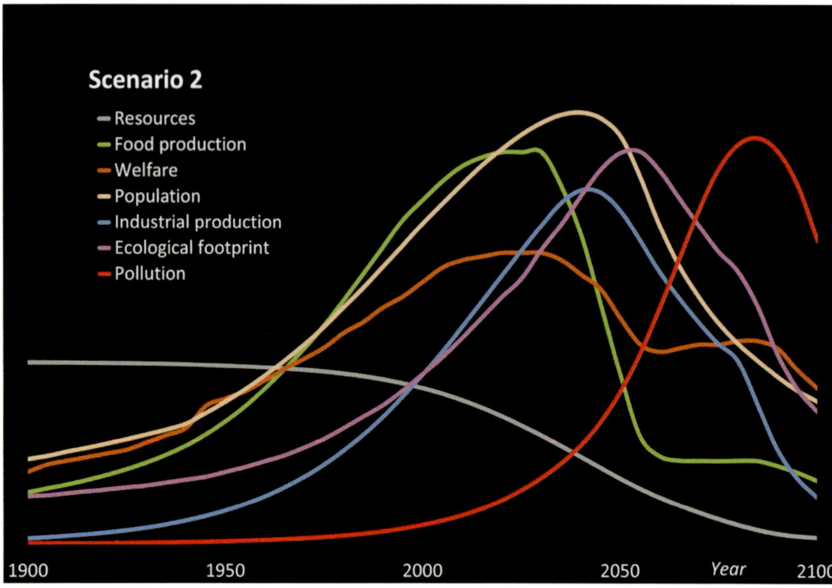

FIGURE 10.4 Meadows et al. (2004), Scenario 2: collapse in the 2nd half of the 20th century.

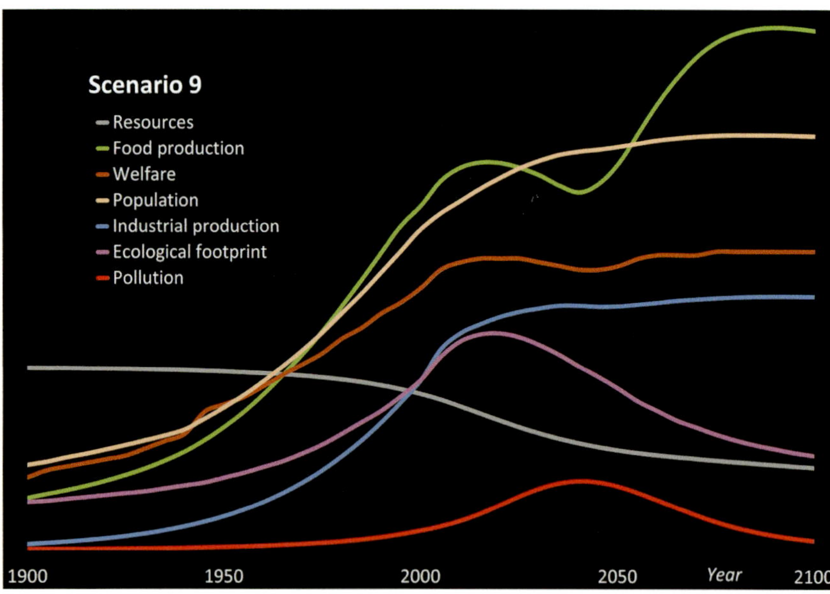

FIGURE 10.5 Meadows et al. (2004), Scenario 9: weakly but still timely inhibited growth.

up and challenged the methods and conclusions with everything they could think of, in an attempt to avoid firm action and thus maintain the short-term profits of the multinational corporations – oil companies, for example. "Garbage", wrote Peter Passell in the *New York Times*. "A PR stunt."[196] It was "a scam", according to Ronald Bailey, science editor of *Forbes* magazine.[197] Then a demonization process broke loose that continues to this day, in the course of which there have even been calls to kill and dismember the authors.[198]

Despite all the sharp criticism, thirty million copies of the book were sold in thirty languages.[199] The fiercely attacked and defended book made it clear to many – but not to all – how all the world's major problems are intertwined.

The book and the underlying computer program represented a profound scientific innovation, as more and more researchers and politicians began to realize in the years following its publication. It proved pointless to keep talking separately about 'the environmental problem', 'the development problem', or 'the resource problem' as if they were separate subjects. In reality, they are all deeply intertwined. That is why we have been talking about 'sustainable development' ever since, because its basic principle is to address all these problems in an integrated way; in a balance between people, planet and profit, as I wrote earlier.

Future scenarios, such as that of Meadows et al., have been run through more often since 1972, including by Meadows' own team in 2004 in 'The 30-year update'.[200] (Figure 10.4 and Figure 10.5 were created using this version. Those who wish to do so, can calculate their own scenarios with the model online.[201])

The picture is still the same: the development can go either way, toward equilibrium (curve C) or collapse (D). The question of the future of both planet and humanity remains open for now.

10.3 First signs

So much for theory. What about reality: are there any signs of collapse yet?

Unfortunately, there are. Earlier in this book it was mentioned that all species of great apes are threatened with extinction. But they are by no means the only species. Although the threat status of only 6% of all (more than two million) animal and plant species has been determined, it is already known that more than thirty thousand species are vulnerable, threatened or seriously endangered, according to the *IUCN Red List* (already mentioned in Chapter 9) in 2021.

The Sixth Extinction

In itself, it is not exceptional for species to disappear. The history of Earthly life is a continuous coming and going of species. This is similar to our own human existence: individual humans are born, live for a while, and then die. Similarly, species are born, live for a while, and then die out. On average, a species exists for between 1 and 10 million years.[202] Of all the species that have inhabited the planet, at least 99% have disappeared, most of them a very long time ago.

But every now and then there have been episodes when the rate was suddenly much higher than normal. Geologists recognize them by fossil material that marks the end of a geological era, such as Permian, Triassic or Cretaceous.

The most famous extinction wave occurred 66 million years ago, at the end of the Cretaceous period. It caused the extinction of 40% of all plant and animal species, including all dinosaurs. There are strong suspicions about the cause: a meteorite impact. The Chicxulub Crater, which lies largely on the sea floor but also partly on Mexico's Yucatan peninsula, is the visible wound.

The most violent extinction wave occurred earlier, however: 252 million years ago, at the end of the Permian. With it, some 80% of all species on the planet disappeared.

Of course, the number of such 'large' extinction waves cannot be determined objectively, because it depends on what you define as 'large'. But many authors identify five such mass extinctions in the geological past. That is not all, however, as several authors wrote books with the ominous title 'The Sixth Extinction'.[203] This Sixth Extinction is happening right now; we are causing it ourselves. Of all mammal species, 26% are currently threatened with extinction, 5% of which are serious. The rate at which mammal species are disappearing is over 2000 times higher than normal. Of all bird species, 14% are threatened, of amphibians 41%, of insects 40%. The details are in Appendix 14.

In 2019, the UN organization Intergovernmental Science-Policy Platform on Biodiversity and Ecosystem Services (IPBES) published a major overview of the state of affairs and reported catastrophic developments. One million species are expected to disappear, with very serious consequences for the well-being of at least 3.2 billion people.[204] In 2021, intensive research showed that only 2.8% of terrestrial ecosystems were still intact; until then, it had been believed that at least 20% had remained.[205]

Pandemics

Natural habitats are shrinking and disappearing. As a result, wildlife species are forced into rapidly growing urban areas. At the same time, poachers and tourists are penetrating ever deeper into the remaining habitats. Intensive trade in exotic animals through unhygienic markets and international smuggling also intensifies contact between animals and humans and increases the likelihood of viruses and other potential germs crossing over to humans, causing new diseases to emerge, *zoonoses*.[206] Large-scale monoculture in agriculture also increases that chance.[207] It is therefore inevitable that the Covid-19 pandemic, which began in late 2019, will be far from the last pandemic of the 21st century.[208] With enormous human and economic consequences, as the years 2020 and 2021 showed.[209]

Climate change

Climate change is no longer just theory. 2018 saw record damage from climate effects, with nine disasters in the U.S. alone causing damage exceeding a billion dollars.[210] The summer of 2019 brought unprecedented heat waves almost everywhere in the Northern Hemisphere, accompanied by giant forest fires in Siberia, Greenland and Alaska, among other places. Six months later, the same thing happened in the Southern Hemisphere. In the Netherlands, it got over 40°C (104°F), which had never happened before. In 2021, research institutes reported that the seven years from 2014 to 2020 were the seven warmest years in measured history.[211]

Rising temperatures have numerous consequences. Not only forest fires, but also changes in the weather: certain areas become drier, others wetter; in general, the weather becomes more erratic, more ferocious and therefore more damaging. Agriculture and food supply are threatened.

Exotic diseases, including dengue, tick-borne encephalitis, Rift Valley fever, Oropouche fever, zika and many others, are advancing from (sub)tropical to previously temperate climate zones as their spreaders migrate: mostly ticks and mosquitoes, such as the Asian tiger mosquito and the dengue mosquito.

Sea level is rising. Using Figure 10.6, you can try to estimate the possible long-term consequences for a number of sea-coastal world cities. Those impacts depend on the rate at which the oceans rise, as the four scenarios outlined (from low to extreme) show.

Wind patterns are changing. Increasingly, in South Asia, the monsoon stays off for months before striking with devastating force, causing widespread crop losses. Tropical storms are increasing in strength and wreaking havoc.

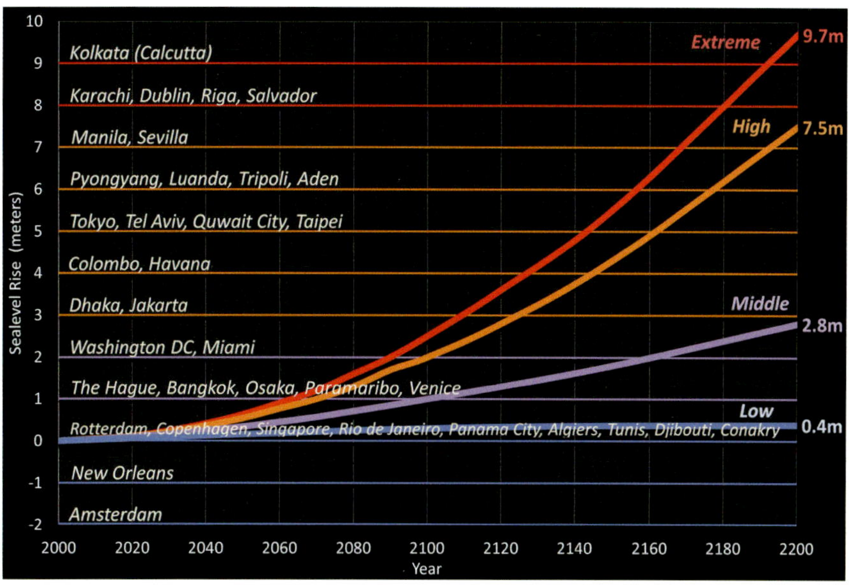

FIGURE 10.6 Sea level rise under four scenarios; and for comparison, the average height above sea level (in 2018) of some cities near seacoasts.[212]

Ocean currents, including the Atlantic Gulf Stream,[213] are weakening, resulting in local climate changes of their own: for example, superstorms in Europe. Reports published by the authoritative Intergovernmental Panel on Climate Change (IPCC) in the summer of 2021 left no trace of doubt: climate change is hitting hard around the world and is bringing us eerily close to nightmarish tipping points.[214]

Tipping points

The greatest danger does not come from slow, creeping developments: some people and countries may still be able to adapt to these. The greatest danger lies in the possibility of sudden, fierce, large-scale catastrophes. These can be the result of tipping points.

Remember Figure 3.1? I repeat it here as Figure 10.7. In it you see some possible tipping points. But those are just examples; there are more.

- *Regarding planet:*
  The extinction of species can cause entire ecosystems to go out of balance in a short period of time, collapse and carry away numerous other species as if in an avalanche.

  Large chunks of land ice can break off and crash into the sea.[215] Or they can slide into the sea as the lower ice layer melts and acts as a sledge. This can cause sea levels to suddenly rise meters. If that happens, you can throw the four slow scenarios of Figure 10.6 into the trash can at once. Among the contenders are the ice of Greenland and the ice shelf of West Antarctica.[216]

  Hypoxic areas in the oceans (in which the oxygen concentration drops) are rapidly increasing in size. *Dead zones* are emerging, which, while not literally dead, are in danger of local extinction, especially for larger animals such as fish.[217]

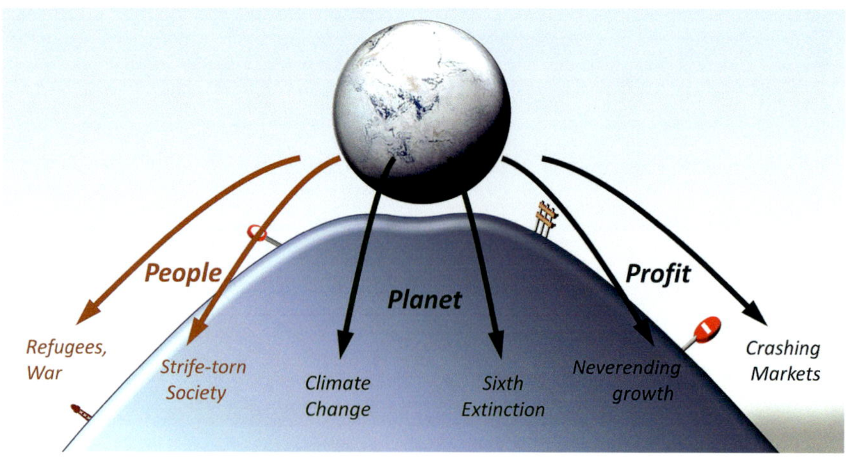

FIGURE 10.7 Abysses in many directions: repetition of Figure 3.1 (page 21).

- *Regarding people:*
  The coronavirus crisis that began in late 2019 was actually a mild pandemic: the Sars-Cov-2 virus was far less contagious and deadly than, say, the Spanish flu of a century earlier. Or the Black Death. The next human-caused pandemic, or the one after that, could easily be much more devastating, plunging both humanity and the economy into a crisis not seen since the Middle Ages.
  When hunger and death strike poor countries, after repeated crop failures due to climate change, inequality between countries takes extreme forms, which can lead to unprecedented flows of refugees that flood rich countries unceasingly.
  Within rich countries, inequality consists of differences not only in wealth and prosperity, but also in health, education, career opportunities and even life expectancy. A study in the Netherlands, published in 2021, showed that low-educated Dutch people die six years earlier than highly educated ones, and live even 15 years shorter in good health.[218] Add to that a housing shortage, plus a new housing crisis (like the one in 2008), angry farmers, anger at immigrants, racism, conspiracy theories, riots, looting, a rock-hard government response, armed terror groups, arson, and the assassination of political figures. The events of the last few years in the U.S., with the storming of the congressional buildings in 2021, and in many countries with Covid riots in the same year, make such a development less improbable than previously seemed. Such an explosive mixture could trigger all-out uprisings or civil wars, perhaps of the caliber of the French Revolution.

- *With respect to profit:*
  Earlier, in Figure 2.3 (page 15), it was shown that the frequency of financial-economic crises is much higher than most people think. This is not from today or yesterday: major financial crises occurred on average every ten years over the past four hundred years.[219]

- *Combinations:*
  There are many more grim threats of tipping points that could lead to collapse in the 21st century. I will keep the enumeration limited here and refer to Appendix 15 for more tipping points and details.
  Each individually, such tipping points are complicated enough. But that is not the whole story, since they naturally influence each other. If one tipping point ignites another, the *mutually reinforcing consequences* of the combination may be considerably more severe than the simple sum of each, making the consequences impossible to quantify. This is true, for example, when an ecological tilt triggers a humanitarian tilt with dramatic, unprecedented consequences. Here is one concrete and unfortunately all too realistic example.

Deadly heat waves

The North China Plain is an area of 310,000 km^2 (112,000 square miles). With a population of 400 million, it is a very densely populated area, which includes the capital Beijing (Peking). It is an important food source for China, both for its own population and for export. Therefore, a lot of irrigation is applied in the previously rather dry area. And it is precisely this irrigation that could well destroy the area.

As a result of climate change, the average temperature in the North China Plain, like in other parts of the world, is rising. This is not good; but, more importantly, the outliers, the peak temperatures that occur only occasionally, are also rising.

Now, high temperatures do occur elsewhere. But they are not always and everywhere equally bad. Their impact is also related to humidity. The more humid the air, the higher the temperature is *perceived* to be. This is because humans regulate their temperature through perspiration: the evaporating sweat dissipates heat and therefore we cool down, so that even in an environment of, say, 50°C or 120°F we can maintain a body temperature of 37°C (98.6°F). But if the air is very humid, little or no evaporation takes place and body temperature is not maintained. Physicists simulate this effect using a mercury thermometer whose liquid reservoir (the glass bulb) is kept constantly wet; hence, the *wet-bulb temperature* measured in this way corresponds well to the *perceived temperature* of the air.

When this perceived temperature exceeds 35°C (95°F), people, even perfectly healthy people, die en masse. This is not unrealistic: in July 2015, the Iranian city of Bandar Mahshahr came awfully close, with a physical temperature of 46°C (115°F) and a humidity of 50%.[220]

In the North China Plain, irrigation, and the resulting evaporation of water, causes humidity to rise. As a result, the perceived temperature is rising faster than the physical temperature. According to researchers, this could lead to peak temperatures of well over 35°C in the latter part of this century – between 2070 and 2100.[221] One can guess what would happen then: panic in cities and towns; the death of millions of people in a matter of days; famine in the rest of China and elsewhere; riots and fighting; and a flood of refugees on an unprecedented scale. What would happen if literally hundreds of millions of people left their homes in agony and swarmed south and west, toward South, Central and West Asia, pushing other peoples ahead of them so that billions of people are adrift, in a massive tsunami all the way to the Atlantic coasts of Western Europe?[222]

A comparison with the infamous European Migration Period, also known as the 'Barbarian Invasions', forces itself upon me. When in the fourth century the Huns, and in the fifth and sixth centuries the Avars marched westward from East Asia, plundering and conquering, West Asian and European peoples were driven from their homes. The Roman Empire collapsed. The Ostrogoths, Visigoths, Alans, Vandals, Suebi, Burgundians, Franks, Alamanni, Angles, Saxons, Lombards, Slavs and Bulgars roamed Europe until they found new settlements or simply disappeared from history. At that time, it was only millions, not billions of people. If something like this happens now: what will NATO do?

The North China Plain is not the only area where such developments could occur. Similar catastrophes are to be expected in South Asia, including highly populated India,[223] and in the already explosive Middle East.

Again: Malthus

Is this the 'revenge of Malthus'? The gloomy cleric was proved wrong, according to many, by the fact that thanks to technical innovations, food production rose much faster than he could ever have expected. (For the details: see Appendix 16.) But it turns out that it is precisely food production technology, in the form of intensive irrigation, that poses a very serious threat to the future of humanity. This is a wry kind of humor.

10.4 The final demise

Many people will intuitively have the idea that our civilization will never really go down. *It won't, will it? It couldn't happen to us.*

Unfortunately, lessons from the past show that it once could. Everyone knows that the once-mighty Roman Empire collapsed; and that is by no means the only empire that has disappeared. As historian and anthropologist Jared Diamond showed in his impressive book *Collapse*, it has happened many times.[224]

> *Just as people are born, live and die;*
> *Just as animal and plant species arise, flourish and disappear;*
> *So do mighty civilizations come, rule and go when the tide goes out.*
> *This is the rule; not the exception.*

Backgrounds for this have been studied not only by Diamond but also by Bas van Bavel, who studied a number of powerful civilizations for which extensive economic data are available.[225] He described how they flourished in years or centuries when society was egalitarian, with relatively small differences in wealth and power between the various classes, partly caused by a consultative structure and a solidarity that ensured that people understood and respected each other's interests. If then, in later centuries, the solidarity in society declined and the inequality in financial wealth and power reached grotesque proportions, society weakened, after which the cracks in the societal fabric, in the cases studied by Van Bavel, led to a decline, a kind of slow collapse over the course of several centuries.

One example is the area of present-day Iraq in the period from 800 to 1400, as Figure 10.8 shows. Note the shape similarity with curve D of the 'Four Growth Curves' (page 130) and with the catastrophic 'Scenario 2' of Meadows et al. (page 134); from the latter graph, the curve of population size has been copied into

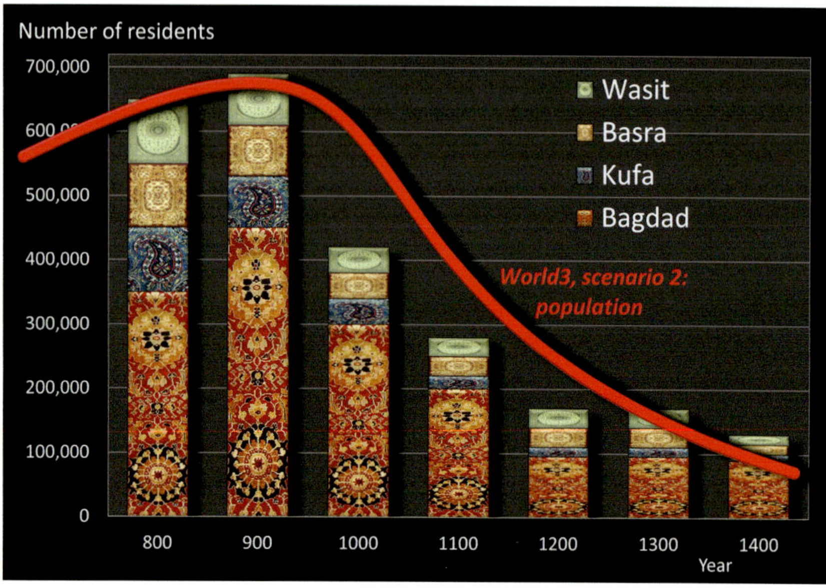

FIGURE 10.8 Iraq, 800–1400: population of the four largest cities.

Figure 10.8. *Collapse:* that's what the graph shows. Over the course of hundreds of years, that's true; but everything today moves very much faster than it used to.

Growing inequality thus poses a great danger to a society. It is an indicator of an impending tipping point in the corner of the Triple P's *people*. Put another way: solidarity, including financial solidarity, is a necessary attribute for the survival of a civilization.

Both Joseph Stiglitz and Walter Scheidel also warned of the dangers of growing inequality.[226] Scheidel noted, using historical evidence, that in prosperous societies growing inequality seems to be a kind of natural and irreversible process in the long run. In his observation, there are only four mechanisms that can reverse this trend, and these 'Four Horsemen' are all tearful: large-scale war; profound revolutions; collapsing states; and catastrophic epidemics.

Does the same apply to us, now? Greer warned in 2005[227]:

> "Any society that exhibits a large increase in capital production, coupled with signs of severe depletion of essential resources, can be considered a potential candidate for catabolic collapse."

And in 2014, Turner compared actual developments over the past forty years with the scenarios of Meadows et al. and found that so far we are faithfully following the life-threatening 'Scenario 2', which is calculated to lead to total collapse during the 21st century.[228]

Of course: the end of the world has been predicted so many times. By preachers of various religions. By environmental thinkers: after Malthus by Paul Ehrlich,[229] among others. And by technology prophets: according to Drexler, the world could soon turn into *gray goo* thanks to nanotechnology, unless Vinge and Kurzweil are right and computers and robots make humans obsolete in one fell swoop via a *singularity** around 2045.[230] Maybe that will save the world, but what kind of world will it be?

Yet the present human civilization still exists, so far. Solutions to problems have always been found, the optimists say, so it will work again. What these optimists forget is that many times things have *not* ended well. The very oldest powerful empire, Babylonia, collapsed long ago. The Persian, Roman, Byzantine, Mongolian and Turkish Empires collapsed. As did that of the Maya, the Incas, and the Aztecs, and this was not exclusively due to the invasion of Europeans, for several Native American civilizations were already severely weakened at that time, in part due to exhaustion of the natural environment.

What is true of individual humans and animals, and equally true of animal and plant species, is also true of the civilizations and powerful empires that dominated a region of the world for a time: *the vast majority came, were successful, and went.*

But: "This time is different," as the already quoted Reinhart & Rogoff ironically titled their book.[231] Is it? Is today's civilization different from all previous ones?

Certainly! For several reasons. On the one hand: it is still there. On the other hand: it is global. Thanks to modern means of communication and transportation, trade, tourism and international organizations, all continents are now tightly connected. Sushi, tortillas, Coca-Cola and cocaine are everywhere. The Internet, YouTube and Facebook are ubiquitous. Climate change is global; new diseases are on all continents within weeks. When stock markets crash, they do so all at once, from Tokyo to Amsterdam to London to New York.

For the first time in world history, if things go badly wrong now, there is no region in the world that is not involved and that can take its place on the world stage after the collapse of a civilization. If we break down now, *everything* breaks down.

This means that the world has no more reserve cultures.

And so, we should hope that the optimists are right and that all the problems are going to be solved. Or better: we should not just hope; we should work very hard at it, as hard as we can, since the problems are not going to solve themselves.

* *Singularity:* the hypothetical point in time when the ever-increasing pace of technological innovations, generated by computers and robots and thus self-reinforcing, gets so out of control that from one moment to the next the innovations explode; so that, faster than the human eye can see, the technology results in something completely unpredictable and incomprehensible to us, after which humans are sidelined.

Will that work? It is certainly not inconceivable. For there is another reason why today's civilization is unique. We know an incredible amount more than we used to, and we can also do much more.

On the one hand, we can therefore destroy more, even on a global scale. And consume more: never before has global biocapacity been exceeded by the human footprint. Never before has the soil been so efficiently robbed of its treasures.

On the other hand, we have scientific knowledge and technological means that will allow us to turn the corner in the right direction, so that we will transform the disastrous 'Graph D' into 'Graph C' and end up in a harmonious balance. For the first time we know what we are doing.

The 21st century is a unique period in world history. After the population explosion in the 20th century, which is still continuing in the current one, we are now in the middle of a technological innovation explosion whose pace, especially thanks to IT, is faster than ever. This creates risks, but also opportunities. If we put our shoulders to the wheel, we can create a global society in which many people can live comfortably.

If we don't, we will lose our chance. And that chance is a one-off. At least, that is what astronomer Fred Hoyle (creator of the term 'Big Bang') believed, writing in his 1964 book *Of Men and Galaxies*:

> "It has often been said that, if the human species fails to make a go of it here on the Earth, some other species will take over the running. In the sense of developing intelligence this is not correct. We have or soon will have, exhausted the necessary physical prerequisites so far as this planet is concerned. With coal gone, oil gone, high-grade metallic ores gone, no species however competent can make the long climb from primitive conditions to high-level technology. This is a one-shot affair. If we fail, this planetary system fails so far as intelligence is concerned. The same will be true of other planetary systems. On each of them there will be one chance, and one chance only."

If the present civilization goes down, it will be the last downfall.

Recommended further reading (for details, consult the online Bibliography)

Book: Diamond, Jared (2005): *Collapse. How Societies Choose to Fail or Survive.*

Book: Meadows, D.L., J. Randers, & D. H. Meadows (2004): *Limits to Growth, the 30-year update.*

Journal article: Herrington, Gaya (2021): Update to limits to growth. Comparing theWorld3 model with empirical data.

Book: Oreskes, Naomi & Erik M. Conway (2014): *The Collapse of Western Civilization: A View from the Future.*

Book: Scheidel, Walter (2017): *The great leveler. Violence and the History of Inequality from the Stone Age to the Twenty-First Century.*

Book: Hoyle, Fred (1964): *Of Men and Galaxies.*

11

THE INEVITABLE CONCLUSION

Proto-economics

In this chapter, I complete the examination of the status of economics as a discipline. For that purpose, I pull together the results of the previous chapters, sorted according to the definition of 'protoscience' on page 59. This consists of a series of seven characteristics:

Logosphere:	1. *Struggling schools.* 2. *Impetus words.*
Mythosphere:	3. *Myths.*
Politosphere:	4. *Copied methods.* 5. *Oversimplification.*
Phenosphere:	6. *Lack of empirics.* 7. *Lack of success.*

11.1 Seven characteristics

I briefly go through the seven characteristics one by one. For those interested, I include some links to applicable but previously unreported literature.

Characteristic 1: Struggling schools

There is a multitude of schools, which are in competition about the key points of the field. This shows that the essence of the field has not yet been grasped.

DOI: 10.4324/9781003306269-12

The large number of schools of economics that continually arrive at mutually conflicting narratives and conclusions makes it clear that this is the case in economics, as was shown in Chapter 6 (from p. 58).[232]

Characteristic 2: Impetus words

The prevailing paradigms in those schools are – at least partially – impetus words.

Enormous damage to nature, man and future is caused by the impetus word *externality*. The same goes for the completely misunderstood word *value*.[233] A long list of other economic impetus words has already passed. *Property* is fundamental, since from it are derived dozens of other words, which therefore also have an impetus character, such as:

Money, banks, capital, pay, price, buy and sell, trade, shares, derivatives, profit, tax, GDP, donate, borrow, rent, lease, interest, theft, conquest, wealth, poverty.

Based on these misconceptions, lying words have been coined, as I told you in Chapter 2: *life insurance, compensation* (page 13).

A side note is appropriate here. It is not the case that I am advocating the radical abolition of the concept of property in all its forms. Personally, I am the owner of a number of items which I would not like to lose. The same is true of all, or almost all, other people. As I wrote in Chapter 9: The notion or the emotion *'mine!'* is not only to be found in humans: it exists just as much in other species. Possessiveness, then, is an innate, genetically determined trait. By ignoring or denying it, as has happened for example in communes, you do violence to people and alienate them from themselves. What I advocate is a fundamentally different use of the word 'property', or its replacement by one or more other words, in order to get rid of the absurd consequences of the current use of the word: think of the *Groningen Civil War* (page 49), of the commercial patents on your genes (page 54), of unaffordable medicines, and of the unscrupulous pursuit of profit by countless companies.

Characteristic 3: Myths

Using these impetus words, questions are asked and narratives are told ('myths'), which are confusing, vague, contradictory, paradoxical or simply proven false.

A long series of automatisms and other myths were described in Chapter 7. Chapter 8 added the *Economic Growth Myth* ("We are not doing well when the economy is not growing").[234]

Characteristic 4: Copied methods

Methods and techniques are used that are successful in other fields, but less so in this field.

See again Chapter 7, which showed how the application of mathematical models, successful in the natural sciences, have been copied straightforwardly into econometrics for the sake of unrealistic models.[235]

Characteristic 5: Oversimplification

While doing so, reality is oversimplified.

It is not only those unrealistic mathematical models that oversimplify reality. Perhaps the greatest oversimplification is that of *homo economicus*, who is nothing more than a caricature, a *cartoon*.[236]

Characteristic 6: Lack of empirics

Experiments and other methods of empirical testing are not performed or are performed infrequently; and to the extent that they are, the results are not or poorly reproducible. There is little need for replicability.

Perhaps you remember the DSGE models. These are the flagship of contemporary economic theory, despite the fact that they have very little to do with reality. The worst part is: there appears to be hardly any need for empirical testing of them. In response to this curious fact, I asked the question: *Why are you talking about such a mathematical model at all?*[237]

Characteristic 7: Lack of success

There is no or little predictive power or other form of 'success'.

Scientific success was defined on page 91 in the form of nine success factors, which I will explore for economics.

> Scientific success consists of:
>
> *Intellectual output:*
> a. Order; b. Understanding; c. Theoretical edifice.
>
> *Societal output:*
> d. Predictive power; e. Technology; f. Contributions to society.
>
> *Emotional output:*
> g. Amazement, beauty; h. Insight into who we are; i. New questions.

Intellectual output:

a. **Order**

I described this form of success as bringing order to chaos: in the form of pattern recognition; of fixed relationships of the type "If A, then always B"; of rhythm, rhyme or repetition. Economics has succeeded in discovering some simple patterns, for example, "If scarcity, then price rise." "If money expansion, then inflation." But there are exceptions to all these patterns; sometimes they mysteriously disappear entirely, as when money expansion in financial markets suddenly does not lead to inflation. In other words, sometimes there is a pattern, sometimes there is not.[238]

Pattern recognition in economics goes even more profoundly wrong when, as Chapter 6 showed using supposed rhythmic wave movements, economic patterns are seen that cannot possibly be real.

b. **Understanding**

In Chapter 13, I will deal with what "understanding" is, so what I write about it here must be somewhat preliminary. Nevertheless, one conclusion can already be drawn.

Flexibly adaptable. Fables and fictions. Novelty, fashion, ideology and power: these are the determining factors in the creation of new economic narratives.[239] Together, the numerous schools of economics have created a motley bundle of narratives that vary so much, that for every economic development there is a narrative to be found or tailored. As a result, the likelihood that some of the narratives, at certain times, exhibit meaningful relationships with reality to some degree is indistinguishable from coincidence.

c. **Theoretical edifice**

As such, the field is still in a *"'pre-paradigm' stage of development in which consensus has yet to emerge on basic concepts and theoretical assumptions."*[240] There is no shared, structured canon.[241]

A theoretical edifice is utterly unfeasible until there is a generally accepted solid fundament on which to build.

Societal output:

d. **Predictive power**

The predictive power of economic science fails at crucial moments. When the economy is in danger of collapsing and it is 'make or break', as in the years around 2008, the majority of economists don't see it coming.[242] Perhaps at other times there is a reasonable degree of predictive power – but that is by no means certain – but then that is of secondary importance: it is like engineers designing a TV set that produces no complaints, 'except when it is on'. Leading economists have proved to be champions at "explaining facts ex-post."[243]

e. **Technology**

Physics brought us *steam engines, moon landings, microwave ovens, the internet, internet banking, feature films, 3D printers, cars, twitter, fMRI scans* and much more. Thanks to physics, I can bring you this book.

Medical science brought us *vaccinations, surgery, antibiotics, anesthesia, dental care, vitamins, dietary guidelines, painkillers, hygiene* and an *increase in healthy life expectancy* by decades.

Agronomy brought us *fertilizers, crop improvement, pest control* and *super harvests.*

These are the practical fruits of accumulated knowledge in certain fields.

Economics brought us *money, banks, stock markets and stocks, mortgages, interest rates, Credit Default Swaps (CDS) and other derivatives*. This may seem like a fine harvest, but these are not the fruits of economics as a scientific discipline. For they all originated in practice, and even much earlier than the beginning of economics as a discipline (usually positioned at Adam Smith, 1723–1790). Look:

Money is some forty thousand years old, initially in the form of shells, pebbles or cattle, among other things.

The oldest *bank* in history was the Taula de la Ciutat in Barcelona, from 1401. The oldest bank still in existence is the Banca Monte dei Paschi di Siena (BMPS), founded in 1472.

The first *stock exchange* was that of Amsterdam, founded by the Vereenigde Oostindische Compagnie (VOC) in 1602.[244] The London Stock Exchange followed in 1659.

Mortgages already existed in classical Indian, Jewish, Greek and Roman antiquity. By 1190, mortgage laws were already enshrined in English law.[245] *Interest* is even much older than the first mortgage: at least 5000 years.

I already wrote about the origins of the CDS in Chapter 2. But the CDS is by no means the oldest financial *derivative*, because that originated around 600 BC by the Greek philosopher Thales of Miletus, as recorded later by Aristotle: he used it to invest in olive oil.[246]

So, all these products are not the fruits of an economic science.

f. **Contributions to society**

The economy has brought prosperity to a minority of the world's population; and even extreme wealth to a few. This can be seen as a success, but because this wealth is not or insufficiently shared, the result is also a gigantic inequality. At the same time, the economic system causes: destruction of the environment and nature. Stock market crashes. Real chances of collapse, perhaps into many abysses at once. The risk of global catastrophes.

Of course, you can list such negative successes in other disciplines as well. Physics brought the nuclear bomb; agronomy brought monoculture. There is no objective method for weighing the pros and cons of a discipline, but I estimate the balance for physics, agronomy and most, if not all, other sciences to be extremely positive. For economics, I cannot draw that conclusion.

Another success, attributed to economics, is the reduction in the number of wars since World War II: there has been talk of the *'Long Peace'*.[247] This is said to be due, in addition to the nuclear threat of the Cold War, to the great interdependence and shared interests of the economies of all countries and regions.

In part, the number of wars has indeed decreased: the modern history of Europe is a successful example of this, as there has never been a war between member states of the European Union, which is unique in the extremely bloody history of the European continent. But it is not clear to what extent this is an economic merit. Moreover, the evidence is nowhere near strong enough statistically: for that, the 'Long Peace' would have to last at least 100 to 140 more years.[248]

Emotional output:

g. **Amazement, beauty**

When I look into the eyes of a gorilla in a zoo, I feel moved, because I know I am near a relative. And at the same time, I feel sadness and helplessness, because I cannot stop the danger that my relative, as a species, is in. Then, as I wander further through the zoo, I experience a deep sense of wonder at the richness of species I see there, which I know is only a minute fraction of the overwhelming species diversity in the world, which did not arise in a Creator's one-time action in a day or six, but through an almost infinite complex of weaving patterns in which species come and go in response to each other thanks to a process of spontaneous evolution that has brought us to the present world over hundreds of millions of years. It is a beautiful and true narrative.

Then, when I think about how this world, our home Planet Earth, is one among a countless number of other worlds, planets orbiting suns belonging to one of the billions of galaxies that fill the universe: I am filled with awe. And I am lucky, because as a physicist I also see the underlying beauty of the grand physical models, written down in mathematical language, such as Einstein's General Theory of Relativity, which – for those who can read them – look like pure poetry.

This is just a quick glance at the prodigious successes of only two mature sciences. How about economics?

When I read about the economic theories; when I study the underlying mathematical models; and when I then look around in the real world: I do not feel amazement but – I already wrote it – bewilderment; and I can't help thinking: why are you doing this, economists?

The next emotion that comes to my mind is aversion. Perhaps it is aroused by certain awful and often haughty answers from economics to the question: who are we?

h. Insight into who we are

In the course of Chapter 9, I provided a series of answers to that question. To begin with, there was the favorite answer of quite a few economics schools:

> *We are: homo economicus. Units of account in economic models.*

Which answer you embrace, depends on who or what you mean by 'we'; that is, from what reference point you view the world. Here is a set of answers that fit the position of the owners. The bosses.

> *We are: the owners of the planet and everything on it. Superior, barely related to the others.*
>
> *We are: the rich and powerful. The economic capital is ours, we earned it ourselves. Using our economic capital.*
>
> *We are: gods. On our Olympus, we are untouchable, infinitely elevated above the common people, and even further above nature.*

If, on the other hand, you belong to the underclass of powerless, or at best to the middle class, you will give very different answers:

> *We are: pawns in the game of the rich and powerful.*
>
> *We are: consumers. Manipulable profit machines; tradable data.*
>
> *We are: factors of production. Human capital.*
>
> *We are: shrinking. Drops in the ocean.*

Wouldn't it be wonderful, and evoke amazement? If economists, along with other scientists, could help us make true:

> *We are: biological, rational, irrational, emotional, greedy, generous, loving, foolish, wise, playful, poetic, beautiful beings. Living in the midst of our brothers and sisters, animals and plants, in our home: Planet Earth, of which we are a part.*

i. **New questions**

And then, if economists work together with other scientists to develop the broad, interdisciplinary new science of omniconomics, then finally some real answers can be found. First, to important questions such as: how do we make poverty disappear from the world, not momentarily but guaranteed permanently? How do we stop the sixth extinction and climate change? How do we live in harmony with nature?

Once those questions are answered, new and exciting questions will keep arising – as in any science – from the amazing theoretical edifice that has then been constructed. Questions that we cannot even think of at present, because we do not yet have the words for them and cannot tell the narratives.

But that will come later. First, we have to answer the age-old questions. With the help of new words and narratives, which will enable us to formulate new policies that will enable us to achieve the desired effects.

11.2 Economics, magic, religion, protoscience

The definition I proposed in Chapter 6 (page 59) states that a protoscience exists if it possesses "all or most of its characteristics". In the first half of the book, this has been examined for the field of economics, with the conclusion that economics convincingly meets all seven requirements. This demonstrated that the discipline is indeed a protoscience.

It is immediately conceded: this conclusion rests on a definition that I myself drafted for the purposes of this book, and that can be seen as a weakness. But if you believe that the conclusion is thereby unjustified, I challenge you to develop your own definition of 'protoscience' and to examine what conclusions it leads to with respect to the status of economics. I invite you to share your conclusions with me.

Practice versus theory

"Yes, but..." is a common counterargument, "all those failures and threats are the result of the policies of politicians and administrators, not economic theory." This raises the question of whether practice is really based on the models of the theorists.

In part, that question was already answered when I pointed out the statements of Robert Lucas, Ben Bernanke, Jean-Philippe Cotis, and Eugene Fama in Chapter 2 (page 16). Their double hats, as university professor and as director at the Fed (the American central bank), the OECD or elsewhere, prove that theory and practice are inextricably intertwined. Bank directors and finance ministers become professors at renowned scientific institutions, while researchers are not only advisors to governments but all too often also become members of them.

But it goes beyond personal unions. The economic models themselves are also actually and structurally used to shape advice and concrete policies. Fisher described in actual stories in 2017[249] how this is happening in the United States, through advisory bodies created by the government and the Fed, including the Board of Governors and the Federal Open Market Committee (FOMC). Christiano confirmed it[250] and showed how the same thing happens at the European Central Bank, the International Monetary Fund, Sveriges Riksbank, the Bank of Canada and Schweizerische Nationalbank. All of them make extensive use of DSGE and related models, the unrealistic nature of which was shown in Chapter 7 (page 81).

In short, yes: economic theory and practice are inextricably linked. The successes and the failures of practice are entirely attributable to the underlying economic theory.

Innovations

Fortunately, serious attempts are being made from within the discipline of economics to come up with stronger models. For example, *behavioral economists* do not work with the simplified homo economicus but with representations of humans that are as realistic as possible.[251] There is appreciation for this, as evidenced by the Nobel Prize that behavioral economist Richard Thaler received in 2017.

Environmental and *sustainability economists* refuse to see nature as an externality and develop methods to internalize it in the economic domain.[252]

An interesting initiative to make the mathematical approach to economics more realistic and much less simplified is called '*econophysics*'.[253] Admittedly, you can see this as a new form of physicalization, but the economic system is treated as fundamentally complex, and this is an important innovation. *Complexity economics* is a new approach that may appear to be of great value.[254] It is not yet the right time to go into that here; it is a main theme in the second part of the book.

Magic and religion

The fact that economics is a protoscience means that the discipline has one leg in science, but the other is still very much in other forms of human creativity. Partly in philosophy, for example in ethics. Partly in politics. And partly in magical thinking and religion.

Economic theorists are eagercredulous; I wrote in Chapter 6 (page 73) about pattern recognition that leads to 'seeing ghosts'. The long series of *automatisms* of economists exude a kind of magical superstition: as if there is some kind of magic power that will make everything work out, while you don't have to do anything to make it happen. An enormous amount of empirical evidence to the contrary has not diminished belief in the automatisms, just as belief in astrology and divination has not diminished.

The automatisms, along with other myths, belong to a colorful mythology that economists tell each other. In doing so, economic doctrine exhibits religious traits. Another feature of this is the inner conviction of many economists that the economic laws are some kind of universal laws that, although formulated by humans, could not have been otherwise. And thus, that their directives and advice to governments and institutions – though contradictory – should be obeyed.

Is that so? Is economics partly a religion? It is informative to quote Yuval Noah Harari's book *Sapiens*. He writes[255]:

> **Religion** *is*
>
> *"a system of human norms and values that is founded on a belief in a superhuman order. This involves two distinct criteria:*
>
> 1. *Religions hold that there is a superhuman order, which is not the product of human whims or agreements. Professional football is not a religion, because despite its many laws, rites and often bizarre rituals, everyone knows that human beings invented football themselves, and FIFA may at any moment enlarge the size of the goal or cancel the offside rule.*
> 2. *Based on this superhuman order, religion establishes norms and values that it considers binding. Many Westerners today believe in ghosts, fairies and reincarnation, but these beliefs are not a source of moral and behavioural standards. As such, they do not constitute a religion."*

The conclusion is inescapable. Economic doctrine is a mixture of magical, religious, philosophical, political and scientific elements and methods. Indeed: economics is a protoscience.

Recommended further reading (for details, consult the online Bibliography)

Book: Morgan, Mary S. (2012): *The World in the Model: How Economists Work and Think.*
Journal article: Ingrao, Bruna (2018): Models in Economics: Fables, Fictions and Stories.
Book chapter: Róna, Peter (2021): Made with Words.
Book: Thaler, Richard (2015): *Misbehaving. The Making of Behavioral Economics.*

Part 2. The *Plan:*

Omniconomics, the new Science

12

BUYING TIME WITH THE SDGs AND MORE

The first half of the book focused on breaking down: of the scientific status of economics.

After proving that economics is a protoscience, with very serious consequences, in the second half of the book it is time to build up. A path is outlined along which a mature science can be created: omniconomics.

In the current chapter, a first, preparatory step on that path is taken: causing postponement. Buying time.

12.1 The SDGs & the Earth Charter

The first chapters said it already: it has to go deeper. Sustainable development has to dig deeper, because what we are doing so far is creating *glued-on* sustainability, while what we need is *intrinsic* sustainability. For this we need to work in all four spheres: not only in the phenosphere (the effects we achieve) and in the politosphere (the policies we implement); but also in the mythosphere (the Big Narratives), which means we need the new omniconomic science. And definitely in the logosphere (the words), because we need new words – or adjusted old ones – to tell the new stories.

Only: such a scientific transition takes time, a lot of time. And in the meantime, the devastation continues: of nature, of the climate, of human solidarity. We cannot afford that time, because stopping the destructive processes also takes a lot of time; so if we are not careful, we will soon have a beautiful omniconomic science and a hopelessly broken world. And so, it is important to postpone, so that we get more time.

That's exactly what we can achieve with the current, glued-on sustainability.

DOI: 10.4324/9781003306269-14

The Leaking Roof

You're sitting at home on a dark, stormy autumn evening in a cozy warm room with a good book and a nice glass. All is well, until you hear "Drip...!"

You look up and discover that the ceiling has a wet spot. You look down and see that the rug possesses an already much larger wet spot, almost like a pond. The roof is leaking!

You jump up and start acting. That is to say..., what are you actually going to do?

The first thing you're sure to do is mitigate the immediate damage. Hastily you combat the worst effects by spreading towels and placing buckets. (Symptom control: phenosphere.) *Right, that's a relief.*

Then you have some time to think and decide what to do next. Do you dare to go up on the roof during the stormy darkness to investigate the cause? Would you rather wait until tomorrow, or even leave it to a handyman?

Finally, next morning, the handywoman comes and gives a report. Depending on that, you can choose. Is the roof still good for the most part? Then we'll fix it. (A policy decision: politosphere.) *Is it very bad? Then it will be a new roof; while we're at it maybe with solar cells.* (A new narrative: mythosphere.) *Or is the leakage due to subsidence of the fundament? Then there is nothing else to do: you are going to build a whole new house* (new fundament: logosphere) *and then move there.*

Roof replacement, or perhaps a new construction: they all take time, as you don't fix it in an hour. So, in addition to such a drastic solution, for the time being you need an emergency provision using towels and buckets. In short: the best strategy consists of a combination of a short-term solution (fighting symptoms: collecting water) and a long-term solution (removing causes: repairing or renewing).

The SDGs

In 1987, *Our Common Future* was published. In this report, also known as the 'Brundtland Report',[256] a research committee commissioned by the United Nations presented a vision concerning the future of our planet and humanity. The publication turned out to be the starting point of what I called the Big Narrative of the 21st century: sustainable development.

Five years later, the first global conference on sustainable development took place: the United Nations Conference on Environment and Development (UNCED), better known as the *Earth Summit*. The summit agreed on *Agenda 21*,[257] the first globally accepted set of sustainability goals, with themes such as poverty reduction, nature conservation, women's and children's rights, and the role of business.

It was a short-term agenda, succeeded already in 2000 by the Millennium Development Goals (MDGs)[258] and then in 2015 by the Sustainable Development Goals (SDGs), also called *Agenda 2030*, because they set goals to be achieved by 2030.[259] The list of seventeen goals can be found in Appendix 17, along with a sizeable set of concrete targets and measurable indicators.

Both the MDGs and the SDGs are set for a period of fifteen years. That is very short, because sustainable development is a long-term process: at least fifty to a hundred years, probably never ending. The globally agreed goals are therefore not aimed at profound systemic change, but mainly at important but swiftly achievable results: among others, the elimination of poverty and hunger; education for all children; women's rights; sustainable use of land and water; sustainable business.

A set of measurable outcomes has been added to the SDGs. On the one hand, this is fine, because it makes them concrete and accountable. On the other hand, this makes them rather business-like and technical in nature, i.e., not very visionary or inspiring. For policymakers this is quite useful, but in recent years it has created the wrong impression that sustainability = SDGs. As a result, there is a tendency to strive for sustainability by ticking off lists.

Yet that is not entirely a good thing, because the SDGs lack a long-term perspective: a 'dot on the horizon' that can lead to a fundamental transition of global society and economy. In itself, that's not bad: it's not what the SDGs are intended for. But for really sustainable development, and certainly for intrinsic sustainability, more is needed.

The Earth Charter

An excellent solution to this is to supplement the SDGs – and their likely successors after 2030 – with inspired texts. One example is the text from which I quoted in Chapter 5 under the heading '*Man belongs to the Earth*', the speech of the Native American chief Seattle (page 56). Another example is the *Earth Charter*.[260] This declaration came about in the year 2000 with the support of UNESCO and the Dutch government; it was signed in the Peace Palace in The Hague. Here is an excerpt from it.

Some texts from the Earth Charter

"We stand at a critical moment in Earth's history, a time when humanity must choose its future. As the world becomes increasingly interdependent and fragile, the future at once holds great peril and great promise. To move forward we must recognize that in the midst of a magnificent diversity of cultures and life forms we are one human family and one Earth community with a common destiny."

"We urgently need a shared vision of basic values to provide an ethical foundation for the emerging world community. Therefore, together in hope we affirm

the risks of certain consequences in time and to take measures. System inertia ('*hysteresis*') can be treacherous, especially when it spans decades or even hundreds of years.

The ozone layer

It was in the 1970s that scientists noticed that a 'hole in the ozone layer' appeared over the Antarctic. It was not literally a hole, but rather a decreasing concentration of ozone in the stratosphere. This was not a good thing, since the ozone layer blocks a lot of harmful ultraviolet solar radiation. Gradually, the ozone layer appeared to be thinning elsewhere as well. The cause turned out to be the breakdown of ozone under the influence of gases such as chlorofluorocarbons (CFCs, 'freon') that were mainly used in spray cans and refrigerators.

It was quite a shock when it thus became clear that humans are capable of causing damage on a global scale with purely small, local causes. That was a setback! First for the scientists. Then for politicians and other policymakers, after they became convinced that the reports were really true. The latter took time, while in the meantime the degradation continued.

When, after a few years, politicians in many countries recognized the problem, international negotiations started – which again took time. Naturally, these negotiations involved not only scientists and environmental clubs but also industrialists who made profits from the production and trade of freon.

On September 16, 1987, the *Montreal Protocol* was signed, which to this day remains one of the most effective global environmental treaties. On January 1, 1989, the Protocol came into force – another year-and-a-half delay – after which the production of CFCs and related substances was banned internationally.

Figure 12.2 shows the effect of that measure.[261] It would take another twenty-five years or so after 1989 for the consumption of CFCs to become virtually zero: the blue line in the graph, the 'cause'. The 'effect' is shown by the red dots: the measurement results of the size of the 'hole' over the South Pole. These show that the size has been slowly decreasing since 2000: the ozone layer is cautiously recovering. (The curved red bar gives an estimate of the trend.) The ozone concentration in the Antarctic is expected to return to, or fall below, its 1989 value sometime between 2050 and 2100.

Addressing the ozone problem involves a long chain of causes and effects, *each of which invokes a delay of at least a few years.*

System inertia in the elimination of ozone depletion

1. *Production and use of ozone-depleting gases (= the cause)*
2. Gradual release of an ultimately significant amount of those gases into the atmosphere

3. Ascending of the gases into the stratosphere (15- to 45-kilometer altitude)
4. Noticeable depletion of the ozone layer
5. Noticeable damage to people and animals
6. *Discovery and proof of the depletion and damage by scientists*
7. *Convincing politicians and other policymakers*
8. *Negotiating a common approach*
9. *Signing a treaty*
10. *Ratifying it by the participating countries, resulting in the treaty coming into force*
11. *Phasing out the production of the gases now banned.*
12. *Removing the gases already in use from existing devices*
13. Reduced concentration of the prohibited gases on the ground
14. Reduced concentration of the prohibited gases in the stratosphere
15. Restoration of the concentration of ozone in the stratosphere (= the solution)

This corresponds to Figure 12.1 *but with fifteen curves!* It is no wonder that this system inertia consisting of many steps makes the recovery take at least a century.

The ozone problem is a relatively easy one, because there were not large national or international economies founded on the use of CFCs: the commercial and geopolitical interests were not very large. And good, easily introduced alternative gases and techniques were available. This is undoubtedly the main reason why the Montreal Protocol became such a successful environmental treaty: it did not cost that much.

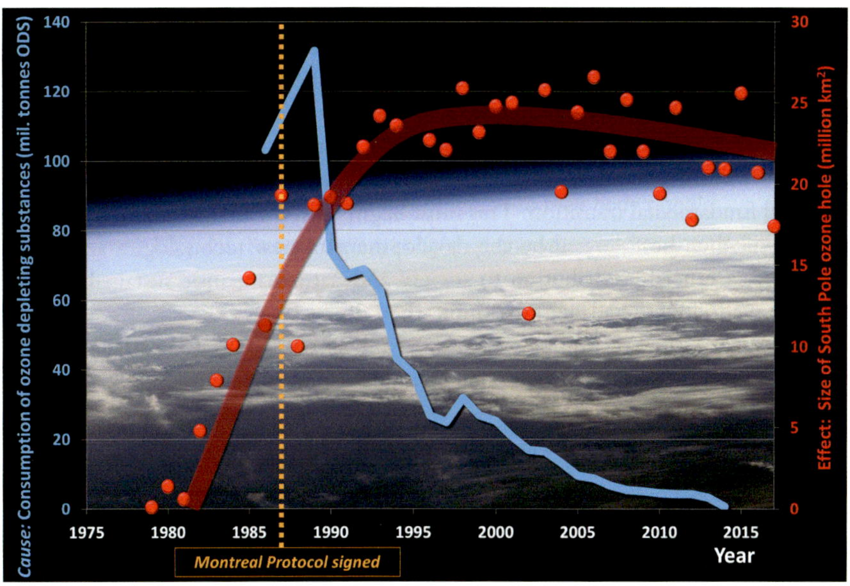

FIGURE 12.2 System inertia: the ozone layer.

reason enough to devote a chapter to them. But to do so, we must first answer the question: what actually is understanding?

Those things

I had no idea what those things were, hovering there pretty much high above my head. It wasn't easy to make sense of it either. Because, at first I just couldn't get the image in focus, so yes, the only thing that came to me was a few vague outlines, which seemed to move back and forth a bit in a rather random way. They weren't even always there.

I was furiously curious about what these phenomena were. And what they were doing there all the time, somewhere above me, with all their strange maneuvers. It seemed like a sort of dance, what those unidentified objects seemed to be doing, flying through the air.

The days ticked by as I studied the unknown objects. Gradually I discovered that I could influence their movement. How I did that, I did not know exactly, but it was unmistakable: I possessed some power over them. If I wanted to, I could order them to move up, down or to the side, until they almost left my field of vision. My field of vision, which day by day gave an ever-clearer picture of them. They were beautiful.

It didn't always go quite right. Sometimes I would instruct them to move in a certain direction, and they would do so, but too violently. Then a sharp pain shot through me, as if I were being hit forcefully by something hard. Then I let out a shrill cry of pain, and lo and behold, something extraordinary happened. I was lifted up. Hugged and kissed. And I got a wonderful satisfaction, when my mouth opened and I could suck. Suck on those soft things. Suck and drink.

That was great! When I let out a cry, then the lifting, cuddling and kissing came naturally. And then came the things with which I could suck and drink. I had power! Not only over those things in the air, which I could make move by simply wanting them to. But also over those other things, those sucking things. I could make those come by calling them. With my voice.

Well, of course I didn't really think that, that I had power. I didn't think at all in words at that age. I couldn't, in those first days and months of my life. But I was curious, extremely curious. Oh, how I wanted to know what it was all about. The world, I mean. I experimented furiously. And with success, because gradually I became handy in making those floating things move in the way I wanted them to. Handy was the right word, because later, when I learned to talk, I started calling them hands. Or feet, depending on the zone where they were mostly. The sucking things turned out to be called breasts.

I had power over my hands, that was more and more certain. I also had power over those breasts, but it was still not easy. My regular method was to summon them with a series of firm cries. But then, that was with varying success. Sometimes they came as soon as I ordered them. But sometimes not, and then I had to scream louder and louder before they finally came. Out of anger I would bite down firmly on them, not realizing that without teeth I could not make such a strong impression.

Sometimes, if I move my hands or feet too wildly, it hurts, and then I have to cry.

Sometimes it's dark and I can't see anything, is everything gone?

Sometimes I can cry out until I'm blue in the face, but the suckling breasts just won't come. Why not? Am I doing something wrong? I'm hungry and thirsty, so I have to keep trying.

Fear. What if the darkness doesn't go away? What if the breasts never come again? Hunger hurts. My hands hurt, too. The world is dangerous, and I don't want it to be. I have power, right? I better figure out exactly how my power works, then I'll be safe.

This is how the experimentation began. What if I use my hands just a little bit more… What if I scream a little louder… or a little bit softer perhaps? Aha: if I pull my face into a smile at the moment that one of those big things hovers above me, then that big thing starts to act very sweetly and to say funny little sounds. Shall I tell you something? If I make those noises myself, then they are even sweeter. So I say, "Da, da!" and "Ma, ma!" Yes, it works! I also have power over these big things. With my voice. With my words. I enchant everything and everyone with my ***words of power****. Everything will be all right.*

The account of my journey of discovery as a baby is a faithful representation of how children and scientists alike learn. Developmental psychologist Rita Kohnstamm wrote the following about newborn babies[262]:

> "Initially, babies do not experience their own body parts as part of themselves. For example: a four-month-old baby can look at his moving hands and be very surprised when, after first waving them aside, they return to his field of vision. He apparently does not yet feel that he is doing this himself. Many babies also bite themselves and then do not feel themselves causing the pain. However, if such body experiences occur regularly in combination (the sensation of swinging a hand, movement before the eyes, the sensation of pain in a hand, the sensation of biting in a mouth) the child learns that it is he who is causing this to happen. A six-month-old baby is no longer surprised by his moving hand. He can even control it deliberately."

The child and the scientist

Someone who starts a new science, behaves exactly as a human child does. Probe. Taste. Bite into it. Try to discover logic, regularity: recognizing patterns. OK, the scientist usually doesn't bite literally, but otherwise the similarity is remarkable. That's why I continue to study what babies and toddlers do, in order to learn how to set up a new science: omniconomics.

As a baby, I too was a true scientist. I experimented passionately. Such as all this screaming and yelling in the crib: the first time I did that, it really wasn't with the intention of luring those breasts. But they did come. And the second time too, and the third. So, at some point I thought (without words, you see), *Hey, that's remarkable! Could it perhaps be…* And so I had a suspicion, a 'hypothesis'. *Could it be that when I scream, those breasts always appear?*

And so: try again, and yeah, there they were again! I had found a pattern. I had discovered a regularity. In natural science this is called a law of nature: *If screaming, then breasts.* I understood!

Soon I discovered more laws of nature. *When hands touch rattle, then funny sound*, hihi! *If smile, then hug.* I embraced every law I could find. I really believed *everything* I thought up. I was *eagercredulous.* I found one truth after another. Not only in the form of if-then laws, but also in the form of concepts. Things. *Words.* Power.

Words gradually became the cores of my natural laws. The nodes, the hard kernels of my understanding, around which my insight into the universe formed. More and more I began to associate the sounds that I initially made automatically with my mouth, such as "Da, da!" and "Ma, ma!", with faces that I saw repeatedly, and I began to call them "Daddy" and "Mommy". From them I learned more words, which showed coherence with each other using those "if-then" phrases. Figure 13.1 shows some of my words and the relationships between them. Together they formed a kind of network.

Needless to say, I did not understand everything I saw or heard. Almost every day new pieces of information came to me, which I then provided with question marks in my rapidly growing personal thought network until I saw a chance to connect them to it.

Thus, I filled my personal logosphere with words. And my thought network became my mythosphere, which in the following years was filled with fairy tales like *Snow White* and (much later) *Life Insurance*, and with myths like *Santa Claus* and *Christ.*

Rita Kohnstamm wrote about this development:

> "… children do not passively absorb their experiences, but immediately actively process them into thinking frames that they employ about features and connections in reality, and that are more than an accumulation of separate experiences. A thinking frame is a way of thinking, a thinking strategy with typical characteristics. That way keeps changing as the child develops."

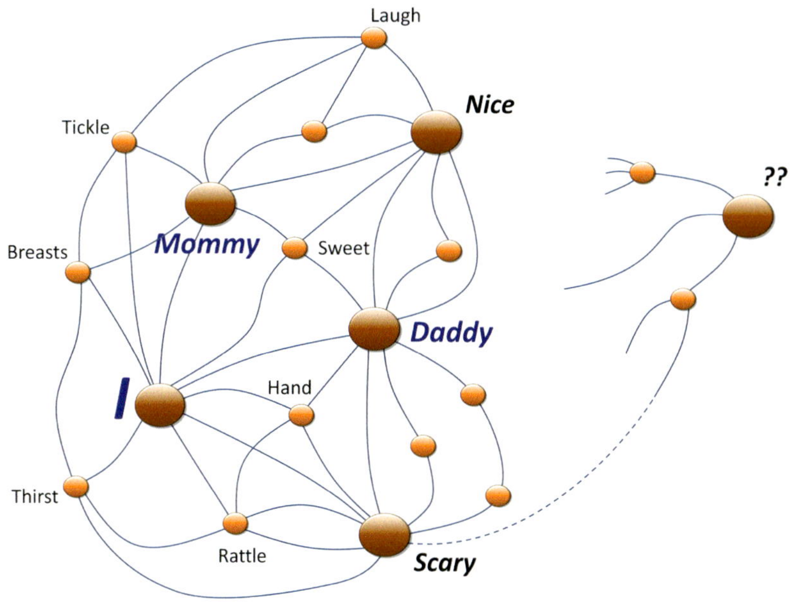

FIGURE 13.1 The thought network of my first concepts.

Law of Conservation of Things

Here is an example. Babies initially have no notion of 'objects'. As they gradually develop that notion, objects at first don't have a permanent existence. The renowned developmental psychologist Jean Piaget wrote about this[263]:

> "The analysis of the first year of mental development shows that object permanence is not innate. The primitive universe during the first months of existence is a universe without objects formed by perceptual images which, as they dissolve into each other, appear and disappear."

Obviously, that was true for me too, when I was a baby. At that time, I didn't take notes; but a generation later I did, after my son Michael was born.

Mommy goes away. And comes back again. Leaves. Comes back. Over and over! That gives a person confidence.

In the nights, the darkness comes. Nothing can be seen anymore. Would it all be gone… and even stay gone? Crying with fear.

Someone puts a handkerchief over my ball. Ball gone! Or over my head. Everything gone! Anxiety. Could it…?

DADDY: Peekaboo! *(pulls away handkerchief).*
MICHAEL: *(laughter of relief)*
DADDY: Peekaboo! *(pulls away handkerchief).*
MICHAEL: *(laughter of relief)*
DADDY: Peek-a-boo! *(etc.)*
1993: Dialogues with my son, 1 year old (he). Minutes made by me.

All those exciting experiments led roughly after about eight months to a new idea: *conservation*. With two laws of nature of its own.

First Law of Conservation of Things: *That which is, stays.*

Second Law of Conservation of Things: *And when it is gone for a moment, it comes back.*

Children love conservation laws, as they give them tremendous security and confidence.

We physicists likewise love both of these laws. In my field, the first law takes, e.g. the form of the *Law of Conservation of Energy*, and we have many more of them. We call the second law *Heisenberg's Uncertainty Principle*: we write it down in the form of mathematical formulas. This also gives us an enormous amount of certainty and confidence.

Religious people, too, are usually very fond of conservation. Many of them cherish a *Law of Conservation of Life (after Death)*: that is a kind of object permanence of the human soul. Talk about security and trust! The common thought among advanced infants, physicists and believers is: surely something that is there, cannot suddenly be no more?

The search for truths

Conservation laws give you the pleasant feeling that you are on to the truth, or at least a piece of it. That you are getting a grip on reality so that it becomes less threatening. That you begin to understand. But is that really the case: is understanding the same as finding the truth?

I am going to investigate this on the basis of two candidate truths: I; and gravity.

Searching truth, candidate 1: I

Naturally, like any toddler, I made the wonderful discovery of that all-important capital of the world: I. Me. Do you have children? If so, did you also enjoy it so much when they were inventing and experimenting with *I*, step by step? Difficult concept, mind you. My son was over 2 years old when he started using it.

> MICHAEL: 'I want to help, Michael wants to help!'
> DADDY: 'Do you like that?'
> MICHAEL: 'Yes, you like that veeeery much!'
> DADDY: 'You are *I*!'
> MICHAEL: 'No, *not* I, Michael Roorda!'
> 1994: Dialogues with my son of 2½ years of age. Minutes made by me.

The examples show that the first use of *I* is a trial and error, just as when you learn to walk. They also show that *I* is not an innate sense but a discovery that you make during childhood.

That is to say… is it a *discovery* or an *invention*?

The sources of the Nile, a television or Snow White

A *discovery* is a thing that was already there, and that you find somewhere during your experiments and quests.

An *invention* is a thing that was not yet there, and that you invent yourself in the course of your quests. You make it, you *create* it. You could patent it.

The sources of the Nile were discovered; they were there before there were people. The television was invented; we didn't find it ready-made somewhere in the bush.

What about *I*: has that always been hidden somewhere deep inside me, from the moment of my conception or a little later, after which, thanks to my active searches, I came across it, *discovered* it? Or is *I* an artificial thing that I constructed myself, just like other people before and after me?

The question is even a little more complicated. The springs of the Nile really exist, and they did before we knew it. Televisions also really exist, but only from the day we invented them.

But we have also invented things that are not real. Superman, for example, and Snow White. Those are fantasy, they're fake. Now suppose *I* was invented by me, not discovered. Is *I* then real, like televisions, or fake, like Snow White? Am I nothing more than fantasy, a fairy tale, a figment? *Did I invent myself, and, if so, am I really there?*

I do think that most people intuitively assume that they are not figments of their imagination. That their self really exists. Maybe you even think I'm asking a stupid question. Come on, if you know one thing for sure!

I think, therefore I exist... or not

That is what René Descartes definitely did: being sure that *I* exists. Or that *I* exist, if you prefer. The reasoning of this 17th-century French philosopher is world-famous. The adult Descartes did the same as every newborn baby does: he examined everything that came into his head. As a first step, he began to wonder if there is anything at all that we know for certain. To test that, he started doubting literally everything. Does France even exist? The sources of the Nile? Do you, reader, exist? Does a heaven and an earth exist? Do I exist?

While doubting, one thing appeared certain to Descartes: there was doubt. Ergo, there was something that doubted. Something that was thinking. That thinking thing calls itself *I*. *I think, therefore I exist. Cogito ergo sum.*[264] And so Descartes proved, according to himself: the existence of *I* is an unquestionable truth. (See Appendix 18 for more details.)

Now, 'unquestionable' literally means: you *can't* question it. But this turned out to be different. For the British philosopher David Hume proved the opposite by indeed questioning the existence of *I*. He concluded that there is no *I* or *self*. However, there is a continual series of impressions that seem to be instinctively connected. From this, people wrongly infer that these impressions together are a continuously existing thing. That thing, I, is in reality a figment of the human mind, according to Hume.[265] A Snow White, so to speak. He did so in line with Buddhism, in which the same view that 'I' is an illusion has existed for millennia. A deception of the mind, *maya*: a fairy tale. (For more on this subject, see Appendix 18.)

According to Descartes, I is an absolute certainty; according to Hume and Buddha, it is an illusion. In any case, the conclusion from this contradiction is, that the 'I' hypothesis has not produced an unquestionable truth. But I go a step further. All things considered, there is not just *one* 'I' hypothesis; one can have a whole range of theories. In Figure 13.2 I have outlined four of them. The first comes closest to the view commonly held in the West and to that of Descartes. The view of Buddha and Hume may resemble the second drawing, upper right, or perhaps the third. I'm not going to explain the four hypotheses further; feel free to come up with your own representation for each. Together they show that the question whether *I* really exist cannot be answered with a simple 'yes' or 'no'. Richard Precht wondered this too, in his book with the wonderful title *Who am I – and if so, how many?*[266]

This is why I wrote, earlier in this book (on page 48), that *I* is an impetus word. A word that leads to contradictions and confusion, especially when related terms are also used, including *free will*, *guilt* and *accountability*.

But: the views expressed so far have all come from religionists and/or philosophers, and they were all people from at least several centuries ago. Fortunately, science has now come to the rescue. Or is it a protoscience that did? I will return to this soon, for first I must urgently raise another issue.

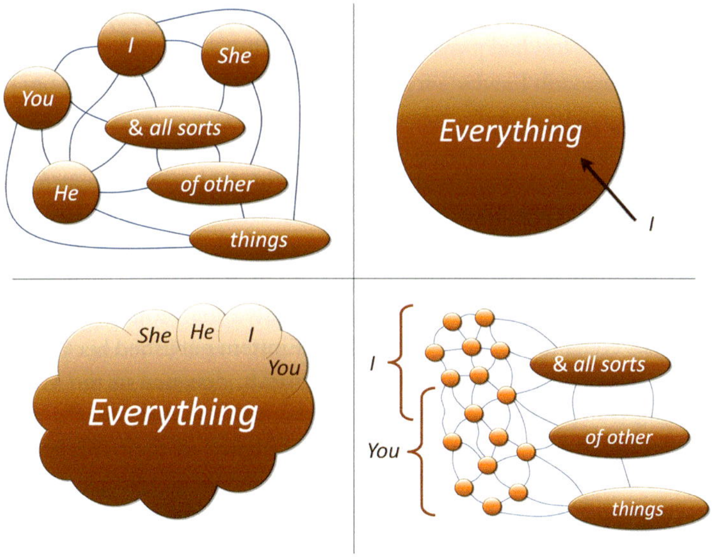

FIGURE 13.2 Theories about me.

Searching truth, candidate 2: gravity

Just as '*I*' is the number 1 success story for children trying to get a grip on their world, *gravity* became that for physicists. As I already wrote on page 45: "I do not make up hypotheses", claimed Isaac Newton, the discoverer of gravity. Or was he the *inventor* of it? Of course: apples were falling from trees long before Newton was born, and planets were faithfully orbiting the sun. Both thanks to gravity, as Newton taught us. But that statement: was it a pure truth?

"Yes indeed," Sir Isaac would have replied firmly if I could have asked him. "For I don't invent hypotheses; I have derived my law of gravity solidly from the observations. All measurements confirm my model!"

He was right in the latter, and physicists enthusiastically agreed for more than two hundred years. During those centuries, the theory of gravity was tested thoroughly. Time after time. At one point an ugly problem arose, when with respect to 'Newton' a minute deviation was found in the orbit of the meanwhile discovered planet Uranus. But that turned out just wonderfully. For the deviation could be neatly explained by assuming that there was an extra, as yet undetected planet somewhere. Using Newton's theory, in 1846 the mathematician Urbain Le Verrier was able, after intensive calculations (without a computer), to predict exactly where this supposed planet should be.[267] Astronomer Johann Gottfried Galle looked at the designated spot with his telescope and that same night he found the planet that would later be named Neptune, the eighth planet of our solar system. The triumph

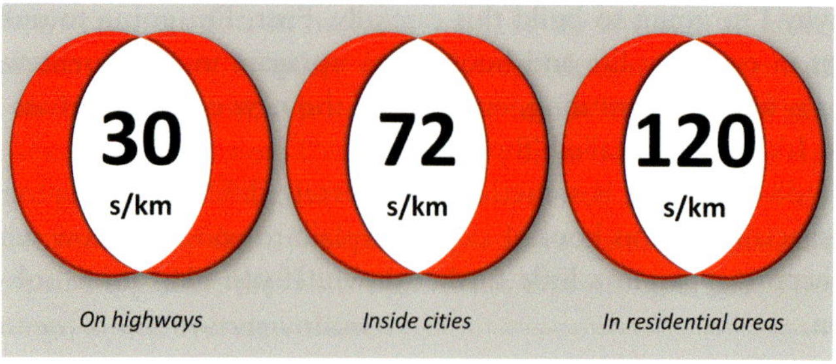

FIGURE 13.5 Minimum sluggity.

kilometer: you can do the math. So instead of a *maximum speed* of 120 km/h, you might as well stipulate a *minimum sluggity* of 30 s/km. If you want to be naughty and go beyond the limits and drive, let's say, 40 s/km, you are free to do so; you will not be fined for it. Look at Figure 13.5 for the new road signs.

Would this minimum sluggity really lead to different traffic behavior? Perhaps the numbers of accidents and traffic violations would decrease. That's good for road safety and for the environment. But bad for the economy, because if the number of accidents decreases, repairs and car sales also decrease and so does GDP. I admit: it is not really likely that it will work. So my proposal is not meant to be serious, more like a way to turn things nicely upside down as a way of brain training.

Upside down: brake energy

I myself have been turned upside down at times. As a teenager I loved science fiction: it stimulated my imagination enormously. For example, in 1969, the year of the first moon landing, as a fourteen-year-old boy I read the book *Chocky* by John Wyndham. It's a fun story, in which the little boy Matthew communicates with an alien lady from an advanced technological society. At one point, Matthew's mother asked him:

> *"What does she find so amusing about it?"*
>
> *Matthew sniffed, paused, and sniffed again.*
>
> *"Pretty nearly everything," he told me, gloomily. "She said the engine is funny, and old-fashioned, and wasteful, and that an engine that needed gears was ridiculous anyway. And that a car that didn't use an engine to stop itself as well as make itself go was stupid.*
>
> *So I told her that's how cars are, anyway."*

A car engine without gears: that I could imagine as a teenager. But then, to use the engine to brake, with the apparent intention of recovering the kinetic energy?

I remember thinking, as a 14-year-old boy: *Yes, it's a nice idea, but you can't capture and store the car's braking energy, can you? You could store it in a flywheel, of course. But that would have to be heavy, and that would cost extra energy during acceleration, more than you save through energy storage. So, it can't be done.*

In short: I did not believe the alien lady. And now look: I drive a hybrid car without gears which does store the energy of motion as soon as I brake. Electrically, in the battery. In a few years – some more saving up – I hope to buy my first fully electric car.

In the meantime, I have had to learn to think the other way around: about a car that not only invests energy to accelerate but also harvests energy by braking. In doing so, I simultaneously learned an even more important lesson: "how things are, anyway", as Matthew put it, does not always mean "how things will be"; and that is an interesting lesson for me, for the economy, and for everyone.

Upside down: keep to the right (or left)

I'll stay with traffic for a moment, to show how great an impact the possession or lack of possession of certain words can have. Imagine if we did not possess the words *left* and *right*. How would you then indicate that we keep to the right in traffic (except in some countries)? For example:

> "Whoever drives in the direction of the north, keep to the east side of the road. Whoever..."

And the rules of right of way?

> "Traffic coming from the southwest has the right of way over..."
>
> "Cyclists with the wind at their backs..., Car drivers with the sun in their eyes..."
>
> "Roll a die..."

Or simply:

> "Whoever drives the fastest..."

Would road traffic as we know it today be possible? In cars, the compass would suddenly be an important navigational tool. What would it mean for road planning and the number of traffic accidents?

The reverse question is actually much more interesting.

What words have we never thought of, that could have made our lives much more enjoyable/ safer/happier in certain ways?

What is missing from our logosphere?

Upside down: the best solution

A few pages ago I discussed the question: what makes apples fall and planets move in circular orbits? Two solutions were proposed for that puzzle: Newton's with the answer *force*, and Einstein's with the answer *curvature*. Newton's solution was mathematically simpler, and that was an advantage, but Einstein's gives better answers, and that ultimately prevailed: Einstein won.

We have become accustomed to always looking for the best solution to puzzles, expecting it to ultimately be the only correct solution, the winning solution. For example, in 19th-century protobiology, the question of the origin of biological species was an interesting puzzle, won by Darwin because his theory of evolution was a better solution than the orthogenetic theory of Chambers and all the other candidates, as I told you in Chapter 6 (page 61).

Now think back for a moment to the cryptogram I showed you on page 177. I give you the solution here.

Cryptogram: The *blue solution*

Horizontal
5. Quality.
7. Clergy.
8. Shiny thing.
10. Accordance.

Vertical
1. Wrong vision.
2. Full of energy.
3. Digital wisdom.
4. Lanthanide below 60.
6. Always a thief...
9. Music over and over again.

1 a		2 b				3 s		4 t
5 c	h	a	r	6 a	c	t	e	r
o		t		i		e		y
7 r	o	t	a		8 h	a	l	o
e		e		9 c		l		u
10 a	g	r	e	e	m	e	n	t
s		y				r		s

The answers add up quite nicely, as you can see. But now look at the following.

Cryptogram: *the red solution*

Horizontal
5. Quality.
7. Clergy.
8. Shiny thing.
10. Accordance.

Vertical
1. Wrong vision.
2. Full of energy.
3. Digital wisdom.
4. Lanthanide below 60.
6. Always a thief...
9. Music over and over again.

1 f		2 s				3 c		4 r
5 a	t	t	r	6 i	b	u	t	e
l		a		t		l		p
7 l	a	m	a		8 o	p	a	l
a		i		9 p		r		a
10 c	o	n	g	r	u	i	t	y
y		a				t		s

That solution is not wrong either. Which is the right solution, the blue or the red? Neither. Or both, just how you want to see it. Because one solution is not better or worse than the other. *There is no winner.*

Now suppose it wasn't about one of those modest little cryptograms, but about something big and important. Let's say: finding a *Theory of Everything*. But *really* everything, that is:

Theory of Really Everything

A scientific model that explains why or how the planets revolve around the Sun; clarifies who or what I am and whether we have free will; teaches us how to avoid the next economic crisis while making everyone rich and saving the environment; and also explains who or what God is or is not.

Moreover, assume that this puzzle has been solved with great success by two different groups: by the *Blue School*, which started from Einstein's theory of relativity, and by the Red School, which chose the Quran and the Bible as its starting point. Suppose that their solutions give exactly the same number of correct answers but differ enormously from one another. Which school then tells us the *truth?*

I admit: finding a *Theory of Really Everything* is a bit trickier than solving a crossword puzzle. The universe is a *very* complicated cryptogram, and the chance that its structure is such that there are really two different, yet equivalent complete *Theories of Really Everything* is probably smaller than the chance of solving a cryptogram twice. Besides, of course, I deliberately designed my cryptogram for that purpose, which wasn't even easy.

Let me fantasize about something else. Suppose that the task of economic protoscience was not to study the flow of money and goods but to solve cryptograms. And suppose that the Marxists, in the left part of my puzzle obviously, had found a piece of the red solution. While the Schumpeterians had at the same time found a piece on the right side, which is natural to them, but had ended up with the *Blue solution*, as shown below. You see immediately: those two schools are never going to agree with each other, that is impossible.

Cryptogram: stuck...

Horizontal
5. Quality.
7. Clergy.
8. Shiny thing.
10. Accordance.

Vertical
1. Wrong vision.
2. Full of energy.
3. Digital wisdom.
4. Lanthanide below 60.
6. Always a thief...
9. Music over and over again.

1 f		2 s				3 s		4 t
5 a	t	t		6		t		r
l		a				e		y
7 l	a	m	a		8 h	a	l	o
a		i		9		l		u
10 c	o	n	g		m	e	n	t
y		a				r		s

Of course, these are all just concoctions. But what about our real world: can the solutions of scientific, ethical, political and other puzzles that we have found so far always be reconciled, that is, fit together smoothly? If not, should that eventually happen, or not? In other words, how large, in principle, is the number of attainable *Theories of Really Everything*: 0, 1 or more than 1?

13.3 Connecting upside down in Economics

Enough practice in everyday life. Let's cut to the chase and take a closer look at economics. That will yield surprising new options.

Upside down: the average income

The words in which you think have an enormous influence on the effects that you get. The following example demonstrates this.

Somewhere in Europe – it could be Asia – lies a very small country: with only three inhabitants. In the year 948 (the country follows its own era), their annual incomes are: 210, 150 and 120 thalers. But it doesn't stay that way, because in the following year the richest inhabitant has managed to increase his income to a whopping 1,200 thalers. That is by no means a bad thing. Unfortunately, the other two citizens don't fare as well, as their incomes drop to 9 and 6 thalers. This is annoying, because one of them just needs fresh clogs and the other a new skirt; they can forget about that for the time being.

The government of this nation naturally keeps track of how the country is doing economically. For that purpose, the widely recognized gross domestic product is used, the GDP. Better yet: the GDP per capita, that is, the average income.

Well: in the year 948, the average income is 210 + 150 + 120 thalers divided by 3, and that is 160 thalers per person. In 949, it comes out to one-third of 1200 + 9 + 6, which is 405 thalers. That represents a handsome increase of over 150%. Proudly the government presents this progress in the national annual report.

Unfortunately, two out of three citizens then pose as somewhat bland, not hesitating to bring up their clothing problems. As if the government had anything to do with that... But at the insistence of the two-headed opposition, the cabinet invites a sage from a neighboring country to give their opinion.

"Look," speaks this one. "You have calculated the average income by adding up the individual incomes. And then dividing them by the number of people." The government nods in agreement. Of course! That's how you do it.

"With that, you have calculated the so-called 'arithmetic' mean income. But something else is possible." The government leans forward curiously.

"Mathematics also offers us," the philosopher continues, "the 'geometric' mean income."

"How does that work?" inquires the government, which (as you might have suspected by now) consists of the richest citizen, who has paid for his influential position entirely by himself.

"For that purpose, you don't add up the incomes," explains the sage, "but you multiply them. You do not divide the result by three but you take the root. The third-power root of course, because three incomes are multiplied."

"Well, why would you do that?" asks the government. "What does it matter anyway?"

"Oh, but it matters a great deal," replies the distinguished guest.

The how and why of this escapes the government. You too, reader? To avoid the risk of you dropping out soon – not everyone loves math – I'll skip the rest of the learned neighbor's argument. Instead, I'll summarize his words and tell you that there is even a third mathematical method for calculating an average, and it's called the 'harmonic' mean.* You can find the mathematical details all in Appendix 20. I'll give you the results here.

(If you want to experiment yourself, you can download a spreadsheet called *Mean incomes – AMI, GMI, HMI* from this book's website.)

Numerical example

	Year 948	*Year 949*	*Year 950*	*Year 951*
Income individual #1	210	1200	1,200,000	400
Income individual #2	150	9	9	400
Income individual #3	120	6	6	400
Arithmetic Mean Income, **AMI**	160	405	400,005	400
Geometric Mean Income, **GMI**	156	40	402	400
Harmonic Mean Income, **HMI**	152	11	11	400

As an aside: the three averages have existed in mathematics for years, but their application to incomes, resulting in the concepts of *AMI*, *GMI*, and *HMI*, is of my imagination: you will not find them in the literature.

The numerical example shows that in the starting year 948 it hardly matters which average you use. The incomes are fairly close together, with the result that *AMI*, the arithmetic mean income, is almost equal to *GMI* and *HMI*, the two other averages.

But now look at the year 949, where much larger income differences have emerged. The *AMI* goes up substantially thanks to the success of Resident 1, with the tragic circumstances of the others being almost entirely irrelevant. But the *GMI*

* Calculate the harmonic mean as follows: take the inverse of each number (i.e., 1 divided by the number); add those inverses; take the inverse of the result again; multiply that by the number of numbers.

falls sharply from last year and the *HMI* much more so: they are visibly more sensitive to inequality than the *AMI*.

In the year that follows, a development takes place that is, of course, too absurd to occur in the real world: resident 1 adds more than a million to his income, while his compatriots do not go forward or backward, can you imagine? The traditional *AMI*, used neatly in the national annual report, goes up proportionally by (almost) a factor of 1000 and the government observes that things are going great. But the *GMI* doesn't go up 1000 but only 10 times, while the *HMI* goes up nothing. That means:

> *The level of the* ***AMI*** *is determined primarily by the wealthiest in a country and virtually ignores inequality and the poverty present. The AMI method is used in real life by practically all countries to calculate economic success.*
>
> *The* ***GMI*** *weighs the poor and the rich roughly equally.*
>
> *The* ***HMI*** *is largely determined by the poor and is insensitive to the increase in wealth of the absurdly rich.*

The *GMI* is always lower than the *AMI*, while the *HMI* is the lowest of all three. This is a mathematical necessity, with one exception: if all amounts are exactly equal. This proves to be the case in the year 951, when the poverty-stricken underclass of our tiny country no longer puts up with it and decides with a two-thirds majority to redistribute the income: from now on everyone gets the same. The result is that the three averages are now all the same, too.

The logical question now is: which average is the *true* average? The right one, the best one?

The answer is: there is no best answer. Each of the three averages is useful, but for different purposes. A government that cares primarily about the welfare of its wealthier citizens chooses the *AMI*, that is, the traditional gross domestic product (GDP). A government policy in which all people matter, could work with the *GMI*, but as far as I know this does not happen anywhere. The same is true of the *HMI*, which a government might choose in order to focus its policy primarily on eliminating poverty and inequality.

You can see here a clear example of how the choice of a word (*AMI*, *GMI* or *HMI*) determines what story is told ("What is progress?"), after which that story determines what political policies are pursued and thus what effects are achieved, such as increases in inequality or, conversely, decreases in poverty: see Figure 13.6.

At least that is the theory, because in practice there is no choice at all, the words are chosen by the richest and most powerful people and corporations, and so the outcome they want determines the narrative and policy. This is why every country uses the *AMI*.

Quite a few economists and politicians will presumably shrug at this because they are convinced that the *AMI* is the only correct, logical, true method of measuring a country's wealth. I can't refer here to a source of literature that has examined

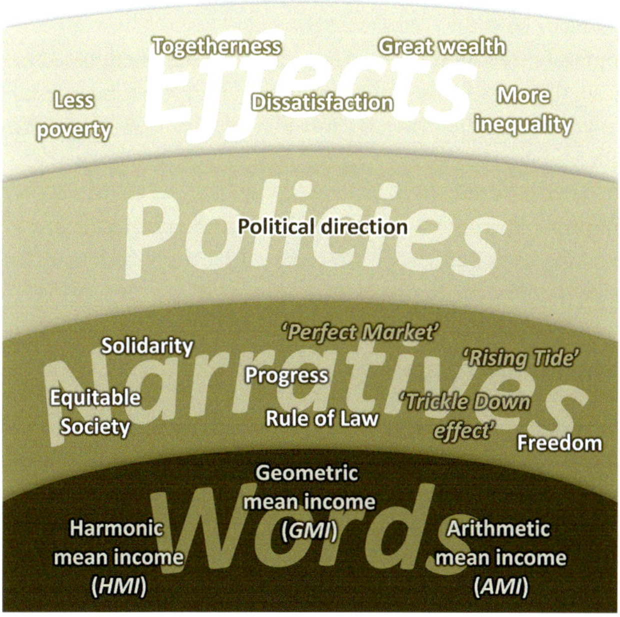

FIGURE 13.6 Four spheres and mean income.

that, because as far as I know such a study has never been conducted. But if that is true, it illustrates an important fact, namely that most economists unconsciously assume that their economic laws are necessary and cannot be changed. Of course, this is not true, because economic laws are made by humans and therefore can be changed by humans. Economic laws are thus fundamentally different from physical laws, as the following experiment shows.

Tape three one-pound weights under a thin loose slat that you hold horizontally. The first weight you hang 210 millimeters from the left side. The second at 150 and the third at 120 millimeters. (These are the same numbers as the incomes in thalers in the year 948.) The center of gravity is now the point at which you can rest the slat on one horizontally extended finger, such that the slat is balanced.

Question: how far from the left side is the center of gravity?

Answer: add up the three positions: 210 + 150 + 120 = 480 mm. Divide the answer by 3 and you get: 160 mm: equal to the AMI in thalers. So, the center of gravity is the arithmetic mean of the positions of the weights.

You can also use the positions of the three weights to calculate a geometric or a harmonic mean. You will then get results, but they do not make any sense, they have no real meaning. The center of gravity is the arithmetic mean, not the geometric or harmonic mean. We are not allowed to choose, there is only one winning answer.

Conclusion: laws of nature are not freely chosen. The dates they describe have been discovered, as have the 'sources of the Nile'.

On the opposite side: political and legal laws are designed, based on self-chosen preferences, so they are freely definable. Since they have been established, they really exist, they are like 'televisions'. Just like televisions, we can change them to suit our needs.

Economic laws have also been invented, but many of them are myths: they are like 'Snow White'. Three such myths are found in Figure 13.6.

The choice of economic laws is determined by the values you cherish. Value-free economics therefore does not exist.

Upside down: Money Conservation Laws

If I pay you a tenner, you are ten euros (/pounds/dollars) richer and I am ten euros poorer. So when we add up, we still own the same amount. This suggests that there is some kind of *Law of Conservation of Money*. This is not surprising, as it is an obvious idea that money is the substance of which the economy is composed, just as energy (or mass, which is the same thing) is the substance of which the universe is composed. Energy is a conserved quantity: it cannot become more or less, because energy is 'stuff'. Money is also stuff, and all stuff, as we discovered in our infancy, is conserved: it does not come into being by itself and it does not disappear by itself. Money possesses – instinctively – object permanence.

And yet that is not true. Money can indeed disappear, and you can do it all by yourself. Task: take a tenner out of your wallet; you can use the tenner I just paid you. Take a lighter and burn the tenner. All the way through. And look: the amount of money in the world has just decreased by ten euros. So: there's no conservation law.

The money supply can also increase spontaneously. If you or I do that, we are called counterfeiters and punished. Burning the tenner is not punishable, at least in the European Union, but we should not go too crazy, because then we are screwed according to the European Commission:

> "Member States should neither prohibit nor penalize total destruction of small quantities of euro banknotes or coins by individuals. They should however prohibit unauthorized destruction of large quantities of euro banknotes or coins."[270]

Individuals may therefore violate the *Law of Conservation of Money* a little in a downward direction; for the rest we must bow our heads in humility and respect money. However, this does not apply to banks. I wrote it before: banks are allowed to create money almost without limit, and that doesn't count as counterfeiting. They do so every time they issue a loan, like a mortgage or an investment loan. Almost all the money in circulation is created by commercial banks. This is allowed; they are not breaking any law with it. When a loan is repaid, the borrowed money disappears again. Not in a vault: in thin air.

If this feels strange and a bit uncomfortable: I already said so in Chapter 2 (on page 12): money is not a material thing at all. You can't eat money and you can't build a house with it. Money is a fairy tale. A Snow White.

Not only in banks is money value created or destroyed, it also happens constantly on the stock markets. Sometimes in a violent way, as on April 23, 2013. On that day, the *Associated Press* Twitter account was hacked, after which a fake message was sent out into the world that the White House had been attacked and that President Obama was injured. Promptly, a *flash crash* broke out, causing a value of $136 billion in the *S&P 500 Index* to 'evaporate' within two minutes because the prices of many stocks suddenly collapsed.[271] A fortune disappeared into nirvana, into the great void. Most of it, by the way, later came back spontaneously from non-existence.

Still, maybe it wouldn't be so crazy if we introduced some more rules that came close to a conservation law, if not of money, then of money flows. Here are a few ideas for new rules: (1) The total wage bill in a country is fixed. Therefore, if someone starts earning more, others will earn less. What political consequences would that have? (2) The minimum income is a fixed percentage of the maximum income. Choose whether that applies within companies and/or in a country. (Or the whole world?)

Instead, you can choose to put aside the very idea of a conservation law. You can do that, for example, by giving money the same status as vouchers: it automatically, legally determined, loses its value after a certain period of time. This can be done all at once after one year, or after ten years, but it can also be done gradually, for instance by depreciating a fixed percentage of the value of all money from the year it was issued. Ten percent per year, for example: if that ten-euro bill was printed in 2015 (and not burned since), it will be worth 9 euros in 2016, 8.10 euros in 2017, and so on. It will take some getting used to, but there is no technical objection to making such a thing happen. It directly solves a problem that banks are currently struggling with: *inflation just won't come*. Inflation is important, say economists, because otherwise people will hoard their money too long and buy too little, causing the economy to stagnate. Journalist Peter de Waard asked during an interview in 2019: "Why is it that we are not succeeding in bringing about more inflation?" The answer came from Groningen professor and former director of the Dutch National Bank, Lex Hoogduin. "Economic science doesn't really know," he said.[272] With our numerous schools of economics, that shouldn't surprise anyone. None of that is a problem anymore, if you include depreciation in the definition of 'money' in advance. We have the liberty to do so, because, after all, money is a fairy tale that we write ourselves.

Upside down: money in 2110, or: what do you believe in?

You know the A$eists, don't you? But you may not know that they have their origins in a small village in Ontario, Canada: Fort Frances, which is right up against the United States. Almost touching Fort Frances, but on the other side of the border in Minnesota, is a village appropriately named International Falls. Of the twin villages, the Canadian one, with a population of 8,000, is larger than its southern neighbor, which has a population of 6,000 and is shrinking.

For many years, trade was thriving between the villages, which is said to have consisted mainly of the smuggling of clean water. The trade was settled in U.S. dollars (USD), and this went well for quite a while, until its value compared to the Canadian dollar (CAD) fell sharply in the first half of 2003. Although in retrospect the drop was not too bad (at the beginning of 2003, USD 1.00 was worth almost CAD 1.60, half a year later CAD 1.35), there were a few disappointed traders who had seen better times and were saying that they had lost faith in the US dollar. Others, hearing that, mused that this dollar no longer existed. Had it perhaps been discontinued?

It only took a few weeks until the story was buzzing around Fort Frances that their southern neighbor's dollar had never existed at all. That it was a figment of the Yankees' imagination. Then, after a preacher from a local church, the New Beginnings Fellowship, confirmed this message from the pulpit while his followers enjoyed Community Soup, the soup was forgotten and A$eism was born.

It didn't stop there. The shocked and passionate A$eists shared their beliefs with anyone who wanted or needed to hear them. The belief that the U.S. dollar did not and had never existed spread like an oil slick. Via highways 11 and 12, the city of Winnipeg was reached and ignited, then Ottawa and Montreal, and as the number of believers (perhaps non-believers is a better word) passed the 30,000 mark, the new church also crossed the ocean, after which especially France, Portugal and Morocco were set ablaze. In major cities, millions of people took to the streets chanting "The dollar doesn't exist!" and "We're being fooled!"

Not long after, the belief that the U.S. dollar was a fairy tale struck such a chord that bank executives and ministers feared the consequences for the global economy. As a result, the dollar's exchange rate dropped precipitously, creating the possibility that the A$eist belief was literally going to become a self-fulfilling prophecy. Anyway, you know what happened next.

Granted: you don't know what happened next, because the above didn't happen. On the other hand: does the dollar actually exist? There's a good chance you're convinced of that to a reasonable degree. But the question you might ask yourself is: if you do indeed believe in the existence of the dollar, *what do you actually believe in?*

If someone from the year 1880 saw our current money, which takes the form of numbers on luminous plastic screens, they would probably not recognize it as money. If we assured them it was real money, what do you think: would they believe us?

What will people believe in, in the year 2110? Will money still exist in that year? Maybe it will, maybe it won't. If it is still there, it will surely be something completely different from today. *If someone from today were to see the money of 2110, they would probably not even recognize it as money.*

Upside down; but not enough

I'll end the upside-down thinking with a remark. For they may be quite nice ideas that you have seen here. But please don't think that these and similar proposals are sufficient to make the global economic system intrinsically sustainable. I have not proposed complete solutions; they are just examples to open your mind. A thorough redesign of the economy will require much, much more.

Recommended further reading (for details, consult the online Bibliography)

Book: Gunster, Berthold (2010): *Omdenken, the Dutch art of flip-thinking.*
Book: Precht, Richard David (2007): *Who Am I and If So How Many? A Journey Through Your Mind.*

14

CHARACTERISTICS OF SCIENCE

14.1 Magic, Religion, Philosophy, Science

The first necessity for such a thoroughly renewed economy is, as I put forward in the first chapters of the book, a new science. Not an economic but an omniconomic science, in which the economic actions of humans are just one of its aspects. For that purpose, it is necessary to start thinking radically differently, for which I have just offered some exercises to loosen the thinking muscles a bit.

In the previous chapter, I wrote about *understanding*, which is the basis of every science and even of the everyday existence of all people: in the cradle, in the classroom and in the street. In doing so, we constantly apply induction, the method that Newton wrote about. We connect dots, we weave individual observations together into ribbons. We *generalize*; that is, we derive general laws from limited data.

Science, then, is not really different from the everyday process of building knowledge and understanding, the process with which babies are engaged in the cradle and students in the classroom. But there is one essential difference: the check.

The check

The check is a regular feature of all good science. Every assumption, every discovered 'law' is rigorously checked. Not just once, but over and over again, even repeated over hundreds of years where possible. When I was a first-year astronomy student, I was lectured by a professor of physics who told the audience that he had spent years of his career making observations for the sole purpose of checking a certain, then 150-year-old group of natural laws. In all those years, he had never found even one small anomaly, he said, adding with a sob in his voice that if he had found one, he might now own a Nobel Prize. (This really happened.)

DOI: 10.4324/9781003306269-16

That check takes place with the help of well-chosen experiments. Where such experiments are not readily possible, the check is made at least by systematically designed observations: think, for example, of cosmology, where most research material is too far away (in kilometers or light years), or of history and archaeology, where the research area is too far away (in years or millennia).

It is precisely the check, the empirical testing, which distinguishes sound scientific research from mundane research. Such mundane research produces conclusions, 'laws', that are effortlessly accepted by the discoverer: not only are babies *eagercredulous*, most adults are as well. Thus, a great many excellent, sensible conclusions arise, and they must, for otherwise humanity would never have survived the past 300,000 years. But quite a few misses also crop up: partly as a result of unjustified generalizations, as a result, partly, of gullibility, and partly of obedience.

"If screaming, then breasts. I do that, I have power. The world revolves around me."

"I take an umbrella with me, so it won't rain."

"If I walk under a ladder, it brings me bad luck."

"The attack on the Twin Towers in 2001 was ordered by the U.S. government."

"If certain poisons are diluted with water and then have a healing effect, extremely diluted substances have an even stronger effect."

"Put a saucer of milk on the doorstep at night, and the brownie will come and do chores for you."

"We believe that through the disobedience of Adam, original sin has spread throughout the human race. It is a corruption of the whole nature and a hereditary defect, with which even the little children are infected in their mother's womb and which produces in man, as from a root, all kinds of sins. It is so ugly and abominable before God, that there is sufficient reason to condemn the human race."

The first sentence is a wordless baby law, an early example of magical thinking. The second and third sentences attest to magical thinking by adults. The next, about the Twin Towers, is one of many popular conspiracy theories. All of these opinions are examples of unjustified generalizations: connections between dotted lines that upon careful inspection turn out to be incorrect.

Magic

The same is true of the comment about diluted poisons, which relates to one of the core beliefs of homeopathy: with the extreme dilution you can build up a desired *potency*. In Appendix 21 you can read how homeopaths even use dilutions where the concentration of the (apparently) active substance is less than 1 molecule to all the water in an entire galaxy. Homeopathy is an astonishing example of the fact that

*"**Fact:** In science, an observation that has been repeatedly confirmed and for all practical purposes is accepted as 'true'. Truth in science, however, is never final, and what is accepted as a fact today may be modified or even discarded tomorrow.*

***Hypothesis:** A tentative statement about the natural world leading to deductions that can be tested. If the deductions are verified, it becomes more probable that the hypothesis is correct. If the deductions are incorrect, the original hypothesis can be abandoned or modified. Hypotheses can be used to build more complex inferences and explanations.*

***Law:** A descriptive generalization about how some aspect of the natural world behaves under stated circumstances.*

***Theory:** In science, a well-substantiated explanation of some aspect of the natural world that can incorporate facts, laws, inferences, and tested hypotheses."*

Taken together, these words conjure up an image of a pure science that deduces hypotheses from facts in an objective, independent manner, that synthesizes hypotheses into laws, and uses them to construct a multifaceted theory, a magnificent edifice that stands like a palace on a rock. Indeed, that was the image that scientists and the general public cherished until at least the middle of the twentieth century. If things went well, the end result would be a magnificent *Theory of Really Everything*.

Magic, religion, philosophy and science: these are four different, but partly overlapping ways of thinking. I call them the *Four Styles of Thinking*. A protoscience, like economics, is a mixture of all four, hopefully on its way to just one of the four: science.

14.2 Philosophy of science

In order to properly assess the expectations regarding the newly to be developed omniconomic science, it is important to first take a good look at what exactly science can, may and does. For it is different from what some people suspect or wish it to be.

Positivism

Pure knowledge: that is what science is supposed to deliver; such was the traditional view. Theoretical science in any case, which was thus rated higher than applied science, where researchers 'get their hands dirty'. As early as the beginning of the 17th century, Francis Bacon, building on ideas from classical Greek antiquity, rated pure mathematics higher than the practical mathematics that in his time was increasingly applied in the 'natural philosophy' of people like Copernicus, Galileo and Kepler.[282] A similar viewpoint, but with regard to science, was expressed rather sharply in the late 19th century:

> "We are tired of seeing our professors degrading their chairs by the pursuit of applied science instead of pure science. We wish for something higher and nobler in this country of mediocrity. Nature calls to us to study her, and our better feelings urge us in the same direction. Young men, looking forward into the world for something to do, see before them this high and noble life, and they see that there is something more honorable than the accumulation of wealth."[283]

Throughout the 20th century, philosophers intensely debated the status and value of science. Quite optimistic and positive judgments were made by almost everyone. This was particularly true of – the name says it all – the positivists, a movement that had been around for a few hundred years and became predominant during the 20th century. The discussions were mainly held within and with a group that called itself the *Wiener Kreis*, or the *Vienna Circle*. Around 1950, the members reached a more or less unanimous consensus on the meaning of science and how it should be practiced. The conclusions were described in what came to be known as the *Received View*.[284] It was a milestone in the philosophy of science.

Appendix 22 describes the conceptions of the *Received View*; see also Appendix 23. Here I summarize the main conclusions in my own words.

Science according to the Received View

Science is an edifice. The fundament consists of a small collection of principles and logical rules. The building blocks are the observations, the hypotheses and the laws. The result is a comprehensive theory. The ultimate goal is to connect all parts and aspects of the universe in the form of one all-encompassing Theory of Really Everything.

Doing science is a continuous process of building one layer on top of another. Science is always advancing.

The different parts of the building must match each other, because the theory must not contain contradictions or paradoxes. Moreover, the structure must correspond to all observations, and this must be continuously monitored. Thanks to that verification, science, and science alone, provides reliable knowledge.

Science is pure and objective. It does not matter who the researchers are or of what gender, philosophy or culture they are, because all will provide the same knowledge. Science makes no moral or ethical judgments and is therefore value-free.

You can see: scientific researchers must be wonderful people. Science builds a kind of Temple, and the scientists are its priests. With their own initiation ritual: the promotion to doctor. Or is science perhaps an art form? According to the National Science Foundation in Washington in 1953, it appears to be.[285]

> "A worker in basic scientific research is motivated by a driving curiosity about the unknown. When his explorations yield new knowledge, he experiences the satisfaction of those who first attain the summit of a mountain or the upper reaches of a river flowing through unmapped territory. Discovery of truth and understanding of nature are his objectives. His professional standing among his fellows depends upon the originality and soundness of his work. Creativeness in science is of a cloth with that of the poet or painter."

Reductionism and other explanatory principles

And then there's another premise, which, although not really part of the *Received View*, is closely related to it. That is the *reductionist* view, which holds that everything greater can be explained from (*reduced to*) smaller building blocks. The behavior of water is explained from the properties of water molecules. We understand engines by studying all their parts. Humans consist of cells, those cells of molecules and those of atoms. If you know the behavior of atoms completely, you can also understand molecules, cells and humans. Even human communities exhibit behavior that can be reduced to that of the individual people who are part of it: sociology and economics are really psychology, according to the reductionist view. Psychological behavior is explained by biological properties: psychology is biology. In the same way, biology is actually chemistry; and chemistry is physics.[286]

Reductionism is an example of an *explanatory principle*, a starting point for devising explanations: in short, a tool for understanding many things. Explanatory principles are among those which form the fundament of the edifice of science, as I have just mentioned. Explanatory principles often come in pairs that are opposites. The opposite of reductionism is *holism*, a way of thinking that tries to understand the meaning of the smaller from the greater of which it is a part. Half a century ago, the holistic view was especially popular in alternative, *New Age*-like circles. Since then, under the influence of the study of complex systems, holism has rapidly become important within science. Omniconomics will certainly be partly holistic in nature.

Another pair of explanatory principles is *causality* versus *finality*, goal-orientedness. With the first term we try to understand how the later arises from the earlier in the form of causes and effects. The second term, finality, inverts the arrow of time, as it were, and tries to see the later as the cause or motive of the earlier. In exact science, finality is not popular and causality is. However, the causal principle sometimes runs into trouble, at least in the eyes of laymen, for example when it is stated by astronomers that there can be no cause of the Big Bang with which the universe started, because at that moment time also started, so there was no 'before'. People sometimes find that difficult to understand, which is not strange when you think about the definition of 'understanding' I gave: everyone is used to the fact that *everything* has a cause. From this habituation, it is easily forgotten that causality is not a logical necessity, but merely a man-made explanatory principle, which apparently cannot be applied everywhere.

In the social sciences, the finality principle is much more popular. That makes sense, because if psychology, for example, investigates people's desires, or business studies investigate the mission of a company, both those investigations are based on the idea of finality, of goals to strive for. This also applies to sustainability science with its 17 sustainability goals, the SDGs.

See also Figure A22.1 in Appendix 22, which shows more explanatory principles.

Cuddle hormone

Exact scientists are not always happy with this. They like to apply the reductionist view; and they claim that any form of finality is actually a consequence of causality at a lower, for example a molecular level. For example, 'love' is explained as a combination of the 'cuddle hormone' *oxytocin*, the bliss-giving neurotransmitter *anandamide*, the satisfying *serotonin*, and a few other brain chemicals. You may remember that I mentioned the discussion about 'free will'. Biologists, arguing from causality, explain to us that we have no free will: they show us that our entire behavior, including our spiritual life, arises from the properties of molecules and electric currents, all of which neatly obey the causal physical and chemical laws. Social scientists generally disagree wholeheartedly with this, and they maintain that we do have free will. They are of course right, when you reason from finality; and that example shows that the proponents and opponents of free will shall probably never agree, because they think from different explanatory principles. They solve the same cryptogram, but from different angles: the fundaments of their scientific edifices do not match.

And economics, what about that? I'll come back to it later; first I want to go through something else. Because the positivist *Received View* has gradually become less popular among scientists and philosophers, and there are good reasons for this.

Falsification

Scientists are humans. The awareness that this has consequences has gradually penetrated into science itself. Scientists, just like everyone else, possess, in addition to all sorts of good qualities, to varying degrees ambition, money hunger, laziness, jealousy, dishonesty and so on. One of these negative characteristics is bias. There are many types of bias; one of them is *confirmation bias*. Not only scientists, but everyone else suffers from it: it means that we tend to pay more attention to facts that agree with our beliefs than to facts that contradict them.

It was philosopher Karl Popper who pointed this out forcefully.[287] When a scientist's task is to verify a theory, they will be primarily focused on observations that confirm the theory. There will then be a tendency to ignore, reinterpret or even modify observations that contradict the theory. Therefore, the task of science is to test every theory, by striving not for *verification* but rather for *falsification*. Only when all sorts of researchers have tried with all their might to undermine a theory, and they have not succeeded, may one conclude that the theory is of some value. Of course, even then you may not declare the theory to be 'true', because it could just happen that, after years or centuries of repeated checks, a theory is disproved.*

This also means, according to Popper, that an assertion (such as a hypothesis, a law, a theory) may only be called 'scientific' if it is in principle falsifiable, i.e. if the assertion really puts something at stake: if ways can be devised by which the assertion might turn out to be false.

Later, Popper even went a step further by stating that if someone calls a non-falsifiable theory scientific, they are guilty of pseudoscience.[288] As a result, non-falsifiable assertions have not necessarily become worthless, but they cannot be viewed as scientific nature. An example may illustrate this.

Before the Big Bang...?

That the Big Bang at the beginning of time would have no cause, is difficult to digest for some. Therefore, a variety of attempts have been made to find a cause anyway.

Some astronomers devised variations on the Big Bang that get around the problem. But so far these do not 'predict' anomalous outcomes of observations; so, they are not falsifiable.[289] According to others, the universe may actually be cyclical: after the initial explosion and the current expansion of the cosmos, there is a period of contraction followed by a 'Big Crunch' where all that exists pops back together, immediately followed by the next Big Bang and so on, in a cosmic breathing without beginning or end.[290] The renowned Stephen Hawking even proposed in his very last publication that the universe known to us is in fact a giant hologram behind which lies a deeper reality.[291] It is conceivable that for some of these theories, ways will be found to make them falsifiable, after which they will be tested as scientific theories.

In addition, certain journals have offered solutions from the fields of philosophy and religion. According to one view, the universe is an intelligent and self-aware being that designed and turned itself on.[292] It has been proposed that the Big Bang is a logical outgrowth of ancient Chinese Taoist principles, whereby yin and yang resulted in time and space.[293] And it was to be expected: it has also been suggested that the Big Bang was initiated by God.[294] While everyone is free to take such ideas seriously, they are not falsifiable.

Popper's touchstone for scientific statements is widely accepted in the scientific community. With it, a breach was made in the positivist stronghold of the *Received View*.

* For example: in 2018 and 2019, two new ways of testing the now over 100-year-old *General Theory of Relativity* were devised and implemented using new technology (Abuter et al., 2018; Amorim et al., 2019). Again, the theory passed the exam.

14.3 Values-based science

Out of the ivory tower

It was not the only breach. Developments in and after World War II made it clear that the fundamental scientists' claim that their field was pure, objective, and value-free could not be sustained. In 1939, Albert Einstein pleaded with U.S. President Roosevelt to develop the atomic bomb, fearing that the Nazis would have it sooner.[295] Thus, Einstein's famous, seemingly innocent formula $E = mc^2$, describing the conversion of mass into energy, decades later led to the horrific deaths of hundreds of thousands of residents of the Japanese cities of Hiroshima and Nagasaki. Physics had lost its innocence.

This had already happened to chemistry earlier, with the introduction of chemical weapons, massively deployed in World War I, and later of pesticides and plastics that damage the environment. And also in biology and medicine, not only in the form of the excesses with experiments on humans by the Nazis, but also with the development of biological weapons, genetic modification and attempts to extend human life to at least 150 years, which (as noted earlier), if successful, will completely destroy the Earth's environment. Or computer science, which may succeed in enabling Artificial Intelligence to supplant humanity and take over rule of the planet.

I'm not even talking about pseudosciences that deliberately seek to cover up harmful effects and use that deception to help commerce. Speaking about commerce, I must unfortunately mention serious science too, which should always be completely independent, but which is, in reality, highly dependent on sponsoring and commercially related donations. Actually, the rule of thumb should be:

Rule of thumb for independent research

Always find out who or what paid for a particular scientific study. If the funders have an interest in a particular outcome, ignore the research completely.

The problem is that if you want to use that rule of thumb consistently, you will have to ignore an unpleasantly high percentage of studies. For example, 75% of clinical trials of drugs in the United States are paid for by the pharmaceutical industry.[296]

In short: the wonderful image of the pure scientist who doesn't get their hands dirty has largely disappeared. It has in part given way to the *Frankenstein scholar*, who creates things that they then don't control. The conclusion is: *value-free science does not exist.*

On the other hand, that is only fortunate. Because where science has led, on the one hand, to enormous growth in consumption and thus to the current, far too large ecological footprint, including climate change and extinct species, the solutions to those problems will also have to come from science. So it is a good thing that the science of sustainability is by definition not value-free, is not practiced 'purely' and neutrally, but rather operates from idealistic drive, just like climate science, medical science, biotechnology, materials science, nanotechnology and more. *Values-based science is desperately needed.*

Economics and omniconomics

What about economics? It is evident that the field of economics is not value-free. Earlier in this book it was shown that the models and recommendations from economic theories have an enormous impact on economic practice and therefore on the lives of all people and on the entire biosphere. Depending on which school of economics prevails, the policies of a country or a company move in one direction or another.

It has also been shown that economic laws seemingly resemble laws of nature, in part because the use of the word 'law' suggests such a similarity. In reality, the laws of economics are chosen; and therefore can be chosen differently. That is to say: are they actually chosen? In reality, they have rather emerged and grown spontaneously, in a kind of autonomous process that nobody could see. The emergence and development of the various human civilizations is not a directed process, it is for a good part simply a process of chance, including the diverse economic systems. In this respect, human culture resembles a fungal culture that also grew spontaneously and erratically: see Figure 14.2.

FIGURE 14.2 Human cultures arose and grew spontaneously in all sorts of places, just like fungal cultures in a Petri dish.

14.4 Aggregation levels

What does an economic system actually consist of? Economist Lionel Robbins knew, when he wrote nearly a century ago, "*Economics is the science which studies human behaviour as a relationship between ends and scarce means which have alternative uses.*"[297] Economics, according to him, is therefore about people and their goods, and that is the answer that will be given by most people.

Yet it is not so simple. In Chapter 6 (page 63) I listed the answers given by the various schools of economics. According to the dominant neoclassical school, the main players in the economy are indeed individual people. The institutional school, as the name suggests, emphasizes institutions such as banks, governments and investment funds. The behaviorists like to add all kinds of other organizations, while the Marxists and the Keynesians focus their attention on social classes.

The remarkable thing is that all these schools look at the same world and yet each sees something different. They do so at different *aggregation levels*; that is, different scales of magnitude. You can see what that means from this short story.

Low-profile visitors

One fine day, Earth was visited by a group of merry extraterrestrial vacationers, who spent a few weeks here with their alien photographic cameras. What did they write in their travelogue, do you think, and with what pictures was that report illuminated?

The answer depends not only on the preferences of the tourists themselves, but also on their ***size****. Some of our travel-loving visitors were about a meter tall, roughly comparable to humans. Their travelogue read:*

> "The dominant species on the Earth is called the 'human'. The human is an animal species of 1.5 to 2 liau in height, endowed with intellect and self-awareness. Humans exert a great influence on their living environment, which they have adapted extensively to their own needs, for example in the form of agricultural areas and cities. Various plant and animal species are intensively used by them and are generally considered their property."

The 'liau' is – trust me on this – the most commonly used unit of length of the visiting aliens, and is equal to their average body height. During the visit, that height varied quite a bit between the groups of visitors – and so did the liau – because some of our guests were not a meter but only a millimeter tall. What do you think: does their travelogue begin like this?

> "The dominant species on the Earth is called the 'human'. The human is an animal species of 1500 to 2000 liau in height, endowed..."

Well, of course not. The report says:

> "The dominant species on the Earth is called the 'ant'. The ant is an animal species of 1.5 to 2 liau in height, endowed with instinct and an unquenchable zeal. Ants exert a great deal of influence on their living environment, which they have adapted extensively to their own needs, for example in the form of agricultural areas and cities. Various plant and animal species are intensively exploited by them, and generally considered their property."

I'll go on for a while. For there were also brave extraterrestrials who were once more a thousand times smaller, that is, a micrometer (a thousandth of a millimeter) in size. The travelogue reads as follows:

> "The dominant species on Earth is called the 'bacterium'. The bacterium is an organism of 1.5 to 2 liau in height, endowed with life and an active ability to respond to stimuli. Bacteria exert a great influence on their habitat, which they have adapted extensively to their own needs, for example, in the form of agricultural areas and cities that take the form of diverse plants and animals (for example, ants and humans) that have become completely dependent on the bacteria."

I'm not there yet. Let me number the steps for convenience. If the human lives at size level 0 (zero), then the ant lives at level −1, a thousand times smaller than the human. The bacteria are then at level −2, a thousand times smaller than the ant. What is at level −3?

> "The dominant species on Earth is called the 'molecule'. The molecule is an object 1.5 to 2 liau high, endowed with diverse chemical abilities. The simplest molecule species consist of only one component; these are also called 'atoms'."

I could go on and on, but the picture is clear, I suspect. What you see is that each aggregation level has its own key players, and they all have their own characteristic features. This plays a role in many ways. For example, water can be wet (namely if it is liquid), while individual water molecules cannot: wetness is simply not a property of them. The sentence 'a water molecule is wet' is neither true nor false; it is meaningless.

Similarly, a landscape or an old house can be 'rustic', while a splinter cannot. A mouse can be 'startled', a bacterium cannot. A human can be lovable or unscrupulous; a gut can't. Each aggregation level has its own distinctive properties. At levels of very small scale, this takes absurd forms, with just about all the properties known to us having disappeared and rather ghostly new ones having taken their place. Tourists of 10^{-15} meters in size, at level −5, therefore wrote rather blurry:

> "The dominant species on Earth are called 'proton' and 'neutron'. Both are 1.5 to 2 liau high, endowed with nuclear capabilities. The neutron possesses no object permanence. Our cameras turned out to be useless, as the objects did not appear on our films or photographs. Even by touch, we could not locate the terrestrial objects, since they don't possess the property of 'location', nor 'speed'. When we actively intervened and 'pinned' them down, we were able to locate them, but by doing so we disturbed the objects to such an extent that the observations say more about us than about them."

The Universal Atlas of Scales

To show you this principle properly, I have added Appendix 24. You are no doubt familiar with the three spatial dimensions: length, width, and height. The fourth dimension is time. To this I add a fifth: scale. The appendix contains an atlas in which you take a journey along the fifth dimension: the *Universal Atlas of Scales*. One page from it is also in Figure 14.3. From page to page in that Atlas, I take steps the size of a factor of 1000 each time, just as I did just now in the narrative.

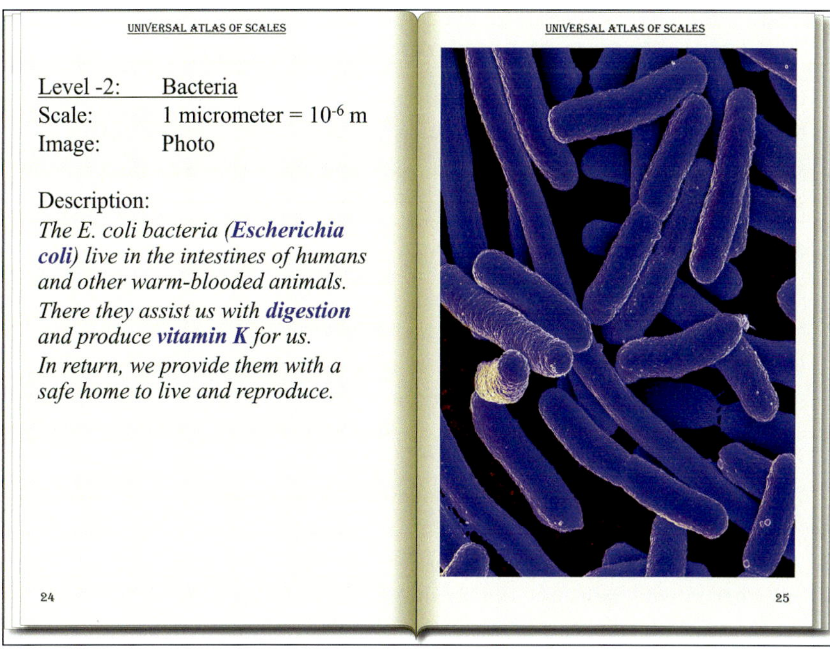

FIGURE 14.3 A page in the Universal Atlas of Scales.

The Atlas starts at aggregation level 9 and ends at level −12; there is no bigger or smaller. We, humans, are near the middle.

And now: on to economics. If you look at the world at level 0, the scale level in the region of a meter, the economy is made up of people. If you look at level 1, you may not see the people anymore but you discover companies, including banks and other institutions. At level 2, you find countries and continents and if you look closely also social classes, and at level 3 our Sun: the *Creator*, the *Sustainer* and ultimately the *Destroyer* of the economy, of Earthly life and of Planet Earth itself. In contrast, if you look at level −3, you will discover the economic significance of nanotechnology.

Which aggregation level is the *right one* for the economics? That question is wrong. Each level has its own relevance.

Why does that matter? Because the picture of the scales, of the aggregation levels, makes it clear that the schools of economics must stop quibbling, as all see only a piece of reality. Together they would see much more. But far from all, as long as the importance of the varying and complementary perspectives of the various scales is not yet adequately recognized and used.[298]

This reflection also makes it clear that economics is not a human science, or at least: not exclusively. Nor is it a social science: it is much more than that – or it should be. Economics as such can never become an independent science, because the economic system is an inseparable part of the omniconomic system.

The research area of omniconomics includes

People, neighborhoods, villages, fields, institutions, businesses, forests, cities, countries, continents, oceans, ecosystems, planets, moons, the Sun and the wind;

and

Goods, raw materials, energy, bacteria, viruses, molecules, atoms, protons, electrons.

And besides:

Love relationships, desires, friendships, rational and irrational behavior, trade relations, infrastructure, internet, history, war, peace, treaties, religion, science, technology, politics, tradition, culture, beauty and delight.

This complex reality calls for fundamentally different thinking, for which so far even the most progressive economists do not seem to be ready. Only when the entire system with all its levels is studied collectively, by many sciences working together interdisciplinarily, can a mature science emerge. All in all, this is an extremely complex whole, so it will not be easy. But modern tools and insights are going to make it possible, as the following chapters will show.

Intrinsic Sustainability Goals (ISGs): a first proposal

It has been established that omniconomics can by no means be a value-free science, but on the contrary should be aimed at the achievement of human and societal goals.

Now the research field of omniconomics has also been defined. That is a good time to make a first, tentative attempt to define the goals of that science as well.

A few chapters ago I gave these long-term goals a name: the *Intrinsic Sustainability Goals* (ISGs). Since those goals relate to our entire, global society, to our house Planet Earth, it goes without saying that they are determined by the collective desires and wisdom of everyone – this will be the subject of the concluding chapter – and therefore certainly not by me alone. But as a fellow inhabitant of our house, I am of course allowed to make some suggestions. I have come up with a few of them. Some of them correspond more or less to the 17 SDGs, some less so.

Proposal for some ISGs

1. **Distribution:** Income and property are shared or distributed fairly so that poverty, hunger and extreme inequality do not exist and everyone can live a bit pleasantly.
2. **Togetherness:** Vulnerable people, such as children, the elderly, the sick, the disabled and those who are *different*, are protected and supported socially and materially. Everyone actually, where necessary.
3. **Labor:** A fair reward is guaranteed for work done. That work is meaningful, appropriate, satisfying and available to all who want it.
4. **Rights:** All human rights are real rights, that is, individually claimable.
5. **Trade:** Fair trade is promoted. Perhaps unfair trade is even impossible or unthinkable.
6. **Environment:** Everyone is naturally stimulated to deal carefully with everything of value, such as raw materials, nature, environment, people, culture, emotions, beauty.
7. **Value:** These values, and those of goods, services and costs, are expressed in a realistic way.
8. **Stability:** Over the course of days, years, decades, and centuries, no major economic fluctuations or catastrophes occur that cause production and trade to stagnate, affecting people and societies.
9. **Participation:** These and other ISGs are established through the active participation of all who wish to do so, aiming at consensus.
10. **Future:** All of this can be sustained indefinitely, for thousands of years, as far as can be reasonably foreseen. Future generations will not be adversely affected by our lifestyle: no mess left behind, no damage, no shortages.

I don't know if you noticed, but in this set of ten characteristics, money is never mentioned, even when it comes to poverty, pay, trade and value. Perhaps intrinsic sustainability can be reconciled with the presence of money. Maybe not. I don't know: it will have to turn out. As said, it may very well be that this future money, if any, would not be recognized as money by anyone of us, right now.

Recommended further reading (for details, consult the online Bibliography)

Book: Popper, Karl (1959): *The Logic of Scientific Discovery*.
Book: West, Geoffrey (2018): *Scale. The Universal Laws of Life, Growth, and Death in Organisms, Cities, and Companies*.

15

THE EIGHTFOLD PATH TO SCIENCE

By defining the field of research of omniconomic science, the starting point of a path that must lead to the realization of that science has been established. The end point of the path has also been determined, at least for the time being, by designating a number of concrete goals, the ISGs, that have yet to be determined more definitely. Time has been created to walk the path: through SDGs, circular economy and other short- to medium-term approaches.

The logical next question is: how do you actually walk on such a path? How do you develop a new science?

I'm going to explore that, based on lessons from the past and the present. Fortunately, the transition from a proto- to a mature science has already been carried out a number of times. It turns out that a fixed series of steps can be distinguished every time. Here is the

Eightfold Path to Science

1. *New physical equipment*
2. *New mental equipment*
3. *New patterns*
4. *New words*
5. *New narrative*
6. *New successes*
7. *New questions*
8. *New fire*

DOI: 10.4324/9781003306269-17

A wonderful example of this in *biology* is Carl Linnaeus' 18th-century classification and naming of all species known to him, which (with a number of corrections) is still in use today. The insights gained from Linnaeus' diagrams were an indispensable link in the establishment of the theory of evolution in the 19th century.

Chemistry offers another brilliant example of a scheme that contributed enormously to the insights of the field. While the *Periodic Table* did not contribute to the breakthrough of chemistry as a science – for that had already taken place – it did contribute to a subsequent breakthrough, in which the basis of chemistry could be established in atomic theory.

By 1800, thanks to improved laboratory techniques, some twenty elements had been discovered, but it did not stop there. The Russian chemist Dmitri Mendeleev knew no less than 62 of them in 1869 (63, he thought, but *dimymium* later turned out not to be an element but a compound substance). Because he also had measurement results of their physical and chemical properties, he saw an opportunity to sort them and put them together in a conveniently arranged scheme.[305] That diagram later became known as the *periodic table of elements*, a pillar of modern chemistry.

It was a sensational diagram, because it was the first time that a pattern appeared in the jumble of thousands of chemical substances. For example, in the diagram the various metals are grouped together; directly below each other you find the ornamental metals copper (Cu, element no. 29), silver (Ag, no. 47) and gold (Au, no. 79), and that feels logical.

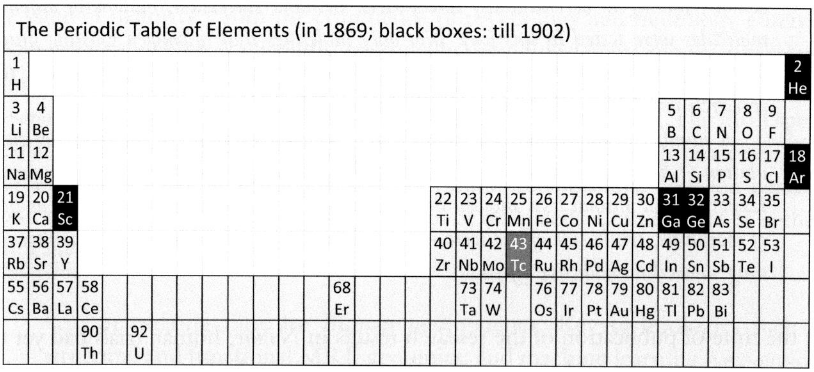

Looking back, one might ask critically: is such an arrangement really essentially different from Johannes Kepler's Cosmic Cup in Chapter 6 (page 66)? Surely that also presented a certain logical-looking arrangement?

The answer is simple. Kepler's cup would have been as much a pillar of contemporary science, *if it had led to successful predictions*. This would have been the case, for example, if the cup had somehow led to the discovery of the planets Uranus and Neptune, which were unknown in Kepler's time. This did not happen; on the contrary, the goblet forbade the discovery of new planets, for the number of six known at the time was crucial.

Mendeleev's system did lead to predictions. For the chemist discovered gaps in his scheme. He predicted that there should be undiscovered elements there, and from their position in the diagram he could even deduce their properties. Thanks to his predictions, the lab technicians knew exactly how to look, and that soon led to success. For in 1875 gallium (Ga, no. 31) was discovered, followed in 1879 by scandium (Sc, no. 21) and in 1886 by germanium (Ge, 32). You can see them in the diagram in black boxes. Unfortunately, element 43 was not to be found, at first. In retrospect, this was no wonder, as it turned out to be an element that decomposes so rapidly radioactively that it has not existed in the wild for a very long time. It was not until 1937 that it was artificially created, after which the element was given the appropriate name technetium (Tc, 43).[306] Earlier, the first noble gases were discovered, helium and argon, with which Mendeleev managed to fill a few other blanks in his system in 1902.[307]

All other empty boxes in the diagram were filled in during the 20th century, the diagram is complete. Just like technetium, all elements from number 93 onwards, including, for example, plutonium (Pu, 94), occur hardly or not at all naturally on Earth, and have been created artificially.

15.4 New words

If, with the use of diagrams, an ordered pattern has been established in the mass of new data, new words – or old words in a new sense – are needed to express the order established. An example is the chemistry word 'element', which was already used in antiquity in the form of the well-known elements *earth*, *water*, *air*, and *fire*, but in the 19th century for the growing group of non-divisible basic substances, of which 62 were known at the time of Mendeleev.

Here is another example. Since ancient times, the prevailing opinion in medicine was that diseases were spread by *bad air*: the name *malaria* even means that literally. People spoke of *miasma*, 'stench', for example coming from putrefaction. In the course of the 19th century, biologists and physicians fought a fierce battle between the supporters of the miasma theory (including the famous Florence Nightingale) and those who believed that diseases were transmitted by *germs* or *pathogens*: living micro-organisms such as *bacteria* and *fungi*. Admittedly, such infections were partly indeed transmissible by air, but partly also by water, food or direct skin contact. Gradually, the pathogen theory won out and, with it, increasing hygiene in hospitals: hand washing and sterile instruments subsequently saved many lives, including through the disappearance of maternal fever. It was not until around 1880 that almost every expert was convinced of the pathogen theory, and miasma had been cast aside as an (in my terms) impetus word. Today, in addition to the germs mentioned, we also know *protists*, the much smaller *viruses*, *viroids* and the very smallest pathogens: *prions*.

It is remarkable to see how much the *Law of Conservation of Belief* applies. For to this day, there are self-appointed 'experts' in 'alternative medicine' who contradict the pathogen theory.[308] Unfortunately, they contribute firmly to a growing hostility to vaccinations.

15.5 New narrative

Schools. All schools, each with its own narrative. This is the characteristic situation for a field of study still in its proto phase. In Chapter 6, I described a series of schools of protobiology in the 18th and 19th centuries, followed by a listing of economic schools, most of which still possess adherents. Such a motley collection of schools may continue to exist as long as a true understanding of the field has not been achieved. Before a generally recognized fundament is constructed, the 'pieces of the puzzle' of the discipline hardly fit together, if at all, as in Figure 15.1, where the 'Yellow School', the 'Green School', and the 'Blue School' can all still shout from their little islands that only they proclaim the truth.

When the time comes that the new tools, patterns and words together powerfully point in one particular direction, then a new story emerges that – presumably after the necessary resistance and perhaps the burning of books or people – brings consensus about a fundamental, comprehensive theory.

The new narrative is the *common, structured canon*[309] of a fresh mature science. Like around 1880 the theory of pathogens, and like earlier Newton's theory of forces (including his gravity hypothesis) and Darwin's theory of evolution. Suddenly, all sorts of puzzle pieces fall into place. The result looks like Figure 15.2: a new science is born.

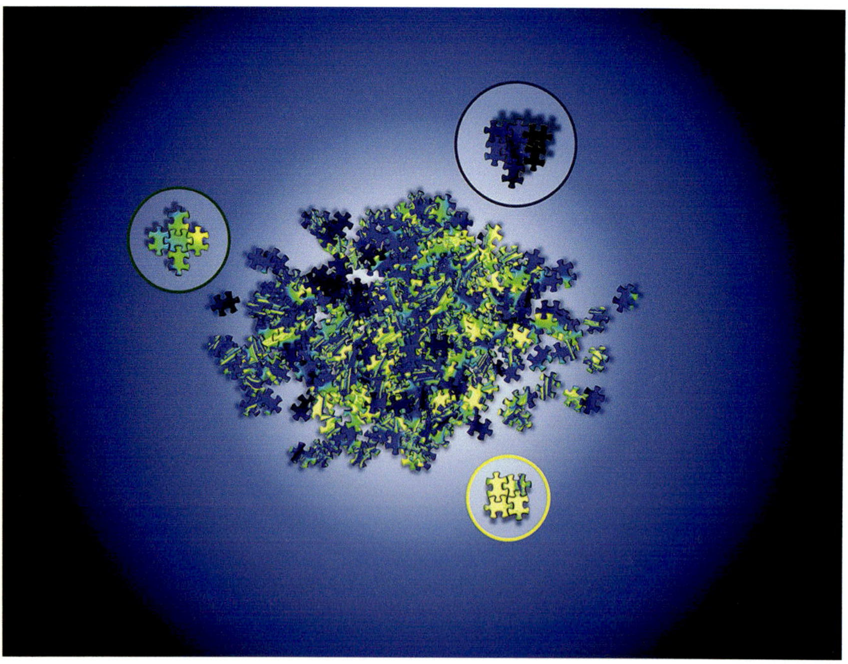

FIGURE 15.1 A discipline in its proto phase: no shared basis yet; struggling schools.

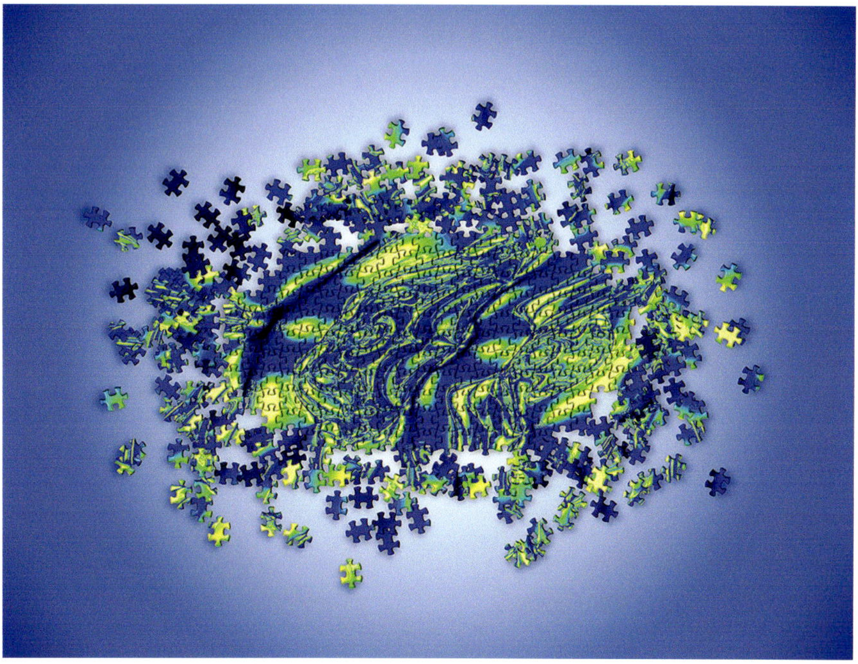

FIGURE 15.2 A young science with a common basis.

15.6 New successes

When in this way all kinds of puzzle pieces fall into place, more success immediately breaks out. Of various kinds, according to the classification with which I opened Chapter 8.

One such success, of course, is: *understanding*. Within the discipline, and subsequently in society, the feeling arises that something is really understood, 'that it fits': that explanations have been found for riddles that have been wrestled with for decades, hundreds or thousands of years. The thought network, like the one in Figure 13.1, is now well-founded and structurally built and tested.

Another form of success is: *technology*. Soon, practically applicable results are emerging, ranging from the steam engine and the digital world to chemical technology, new materials, agricultural techniques, pharmaceuticals, surgery, genetic modification, and weather forecasting. (But also, of course: weapons of mass destruction and climate change.)

Technology is, in fact, an example of *predictive power*. After all, anyone who cobbles together a computer using a blueprint predicts: "I'll put those-and-those parts together so-and-so and the result will work like a computer."

Another kind of success is of an *emotional* nature. Experts and laymen alike look at the fresh theory with amazement and discover a newfound beauty. The theory contributes to an understanding of who we are and what our place in the world is;

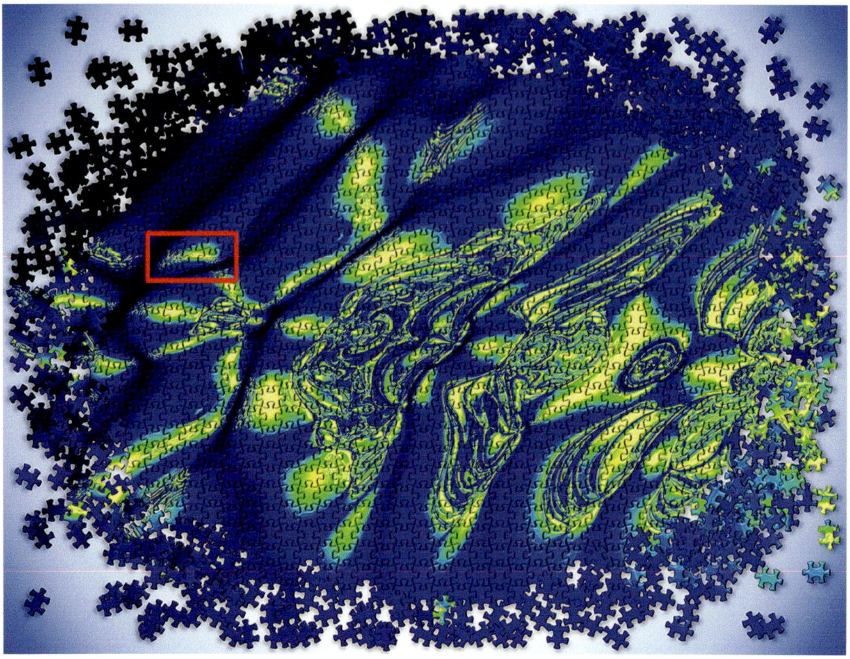

FIGURE 15.3 Advanced Science.

for example, by showing us how our psychic processes work, how we are part of earthly life and evolution, or how old our species, our planet and our universe are.

Following on from that is the *contribution to society*. Technology and new insights allow us to fight poverty, feed the hungry, create freedom, prosperity and luxury, make work more enjoyable, disseminate information, promote communication, encourage and defend emancipation, democracy and human rights.

Needless to say, there is another form of success: *expansion*. Figure 15.2 may have a middle area in which the pieces fit together, but there is also a substantial edge in which they don't yet. Continued successful research expands the solved area, as in Figure 15.3. For example, physics expanded in the 19th century with the field of electromagnetism; and chemistry expanded with the understanding of the elements. Biology in the 20th century entered the path of genetics and the reading of DNA codes. But at the same time, with all these triumphs in every branch of science, the unsolved edge is also constantly growing: there are more loose puzzle pieces in Figure 15.3 than in Figure 15.2. The more you know, the more questions you can ask. A science is never finished.

15.7 New questions

A science is never finished, because each answer raises new questions. Many of these are at the interface between two fields of study or two levels of aggregation.

For example, chemists and biologists combine forces in search of the origin of Earthly life, because the oldest biological evolution must have been preceded by a chemical one, many details of which are not yet understood.[310] The main prize to be won there is: build living matter in the laboratory, using only inanimate chemical materials. Geologists are assisting in the quest by probing the physical and chemical conditions in our planet's oldest past.

Meanwhile, physicists, chemists and technologists are working together to explore the mysterious boundary between the very lowest levels of aggregation in which the laws of quantum mechanics rule and the higher scales of molecules, nano- and microtechnology up to and including the everyday levels where people live and where the 'ordinary' laws of physics apply. How can one understand this curious transition near scale level -3 in the *Universal Atlas of Scales* (Appendix 24)?

Where living organisms are involved, biologists are also scouring for answers to a related fascinating conundrum: the significance of quantum mechanics for life at the very lowest level: that of metabolism and the behavior of chromosomes in living cells. The combined field is called *quantum biology*,[311] and there is also a great deal to discover there.

Photosynthesis

Biologists, chemists, energy technologists and agricultural experts are working closely together to solve a fascinating puzzle that has the potential to have major implications for global society: it is the demand for more efficient photosynthesis,[312] the process by which plants and algae use sunlight with the help of chlorophyll to convert carbon dioxide and water into organic molecules and oxygen. In Earth's living world, photosynthesis is bizarrely inefficient: only 3 to 6% of the sunlight falling on the plant is utilized,[313] or, according to other calculations, even much less.[314] There are several reasons for this, stemming from the fact that the evolution of green plants, dominated by chance, has not exactly produced the smartest chemical processes. The plants absorb only a small part of the sunlight: the rest is reflected back and thus not utilized. You can check that for yourself, because if plants could absorb 100% of the sunlight, they would not be *green* but *black*, perhaps as black as an astronomical black hole from which no light can escape. (Plants were then invisible…)

Another cause is that the enzyme *rubisco*, which is essential for photosynthesis in plants, is not doing so well: in about a quarter of the chemical reactions, not a carbon dioxide molecule is bound but, accidentally, an oxygen molecule, creating a harmful substance that the plant must break down again at great expense of energy. This remarkable and unnecessary mistake probably occurred when photosynthesis was first developed in nature. This was at a time when the Earth's atmosphere contained almost no oxygen and so the problem was virtually meaningless. Today, the atmosphere is 20% oxygen, but plant life has been unable to discover a way to replace the rubisco reaction with more efficient photosynthesis.

Agricultural experts are now trying to create food crops that do not contain this error, with which they hope to increase food yields by as much as 40%. At the same time, chemists and engineers are working hard on artificial photosynthesis, not in plants but in solar cells, which should make it possible to achieve much higher efficiency than that of the current generation of solar cells.

15.8 New fire

If a scientific theory runs into trouble as a result of new observations, it is not immediately discarded. First, researchers see if they can solve the problems with modest additions or corrections. One example was seen in Chapter 13, when Newton's theory of gravity ran into trouble: the orbit of the planet Uranus was incorrect (page 173). As a solution, an undiscovered planet was proposed and actually found in 1846: Neptune. Newton's theory had been saved.

Later, however, it failed to explain a deviation in the orbit of Mercury in such a similar way (see page 174), after which Einstein came up with a brand-new theory that solved the problem.

What would have happened if Einstein had not? According to philosopher of science Imre Lakatos, all sorts of other solutions, from cosmic dust clouds to magnetic fields or whatever, would surely have been tried in order to keep Newton's theory afloat.[315] After all, you don't just give up on a successful theory if you don't have a replacement.

That is a risk, however, because if you are not careful you create on top of a brilliant theory a whole collection of extras, 'stopgaps', all designed to save the beloved theory. A certain theoretical flexibility is fine, but if you take it too far you turn a good theory into a kind of chewing gum: it always fits and so you don't really say anything anymore.[316] Perhaps you remember the flexible application of the set of economic fluctuations, of which I proved experimentally in Chapter 6 (page 70) that it explains everything, and thus nothing.

Stopgaps no longer suffice when the theoretical problems get too big. For instance, physicists and astronomers would love to design a real *Theory of Everything* for physics, but they don't yet succeed. For about a hundred years they have had two powerful structures: Einstein's general theory of relativity, and quantum mechanics. The first is mainly applicable on a cosmic scale, when enormous masses are involved and gigantic distances and timescales. The second is about the very smallest, so about molecules, atoms, elementary particles and quarks. You would say: these two theories never meet, but that is not so. In certain extreme circumstances they are both needed at the same time. You can see one such situation in Figure 15.4, which involves the exciting region near a black hole. Another is the condition immediately after the Big Bang. The problem is: the two theories do not tolerate each other. They can't be made to fit together – until now.

And so, over the course of a century, all sorts of solutions have been enthusiastically put forward. Since the problem cannot be solved with some sort of stopgap,

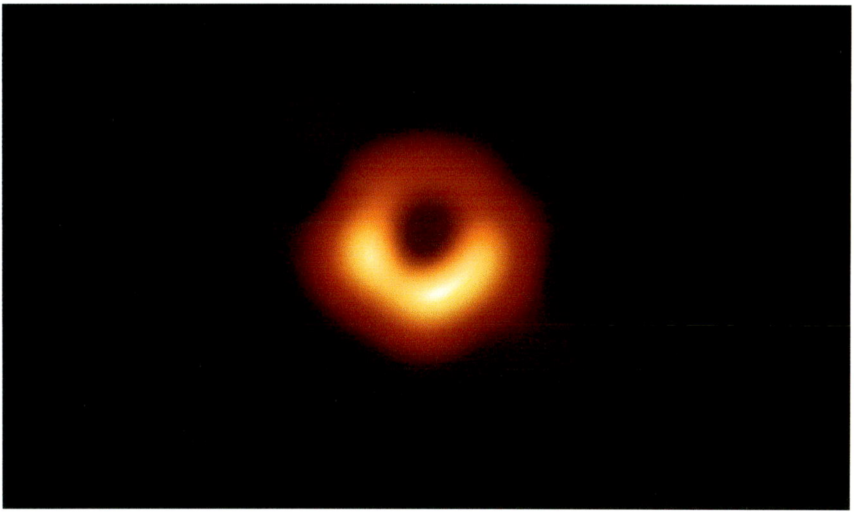

FIGURE 15.4 The first photo of a black hole (2019). The 'hole' is the dark part within the luminescent ring. The existence of black holes was theoretically predicted 103 years earlier.[317]

the creativity of physicists has been greatly stimulated. All sorts of revolutionary ideas have been proposed, from *string theories* to *holograms* and *entropic gravitation.*[318] The battle is fierce, emotions run high at times, and it will continue for some time, since opportunities for empirical testing are scarce and thus there is no referee yet to determine who is right.

Superficially, the situation – conflicting schools in modern physics – resembles the plurality of schools in economics. In reality, however, the situation is fundamentally different, just as Figure 15.3 (advanced science) is very different from Figure 15.1 (protoscience), for scientific contention takes place at the edge of an extensive area of agreement, the *canon*, whereas protoscientific contention does not yet have such an area of agreement at all. *It is precisely this, which makes the difference between proto- and mature science.*

The fire of such a contention in an advanced field of science is wonderful; it produces enormous spiritual energy and free thinkers. A science without such disputes is asleep.[319]

The revolution

If one of the brand-new theories proves to win over the established order, then – as described by Thomas Kuhn – the revolution breaks out.[320] Old key words, the paradigms that prevailed until then, make way for new ones. That is what happened when Newton's paradigm *force* was replaced by Einstein's *curvature* as an explanation for the shapes of planetary and ball orbits.

I outlined the positivist perspective of the *Received View* with the words: *Doing science is a continuous process of building one layer on top of another. Science is always advancing.*

Now, the Received View has been undermined before, for example by the impossibility of objective and value-free science. The occurrence of attempts to 'save' the prevailing theories with stopgaps also does the image of noble science no good but is still justifiable. But scientific revolutions, which demolish part of the majestic edifice? That is an entirely different story. Science is practiced in a much more complex way than the members of the *Wiener Kreis* assumed around 1950.

And yet, even this revolutionary picture must be put into perspective. For a revolution in a *mature science* works out differently from a revolution in a *protoscience*.

Take, for example: the cosmic theory of Ptolemy. The Greek Claudius Ptolemy lived around the year 150 in Alexandria, where he summarized the knowledge of his time in a model in which all the planets and the Sun and Moon revolved around the Earth in circular orbits. (Rather, in epicycles, circles whose center performed a circular orbit around the Earth.)[321] But in the Renaissance, the proverbial *Copernican Revolution* broke out when Nicolaus Copernicus moved the center of the cosmos to the Sun.[322] The scientific, sometimes (thanks to the Catholic Church) deadly battle that then erupted was not finally settled until Newton published his theory a century and a half later. Nowadays, nobody uses the Ptolemaic worldview anymore, except by way of history lessons. 'Ptolemy' has been abolished and forgotten.

Compare that to the later *Einsteinian Revolution*. With his *General Theory of Relativity*, Einstein warped the universe, as symbolically depicted in Figure 15.5. You might expect that 'Newton', outdated as the theory is, would have disappeared through the side door, just like 'Ptolemy'. But that is not the case. In the years I taught physics in high school and college, I fervently explained Newton's gravity to my students and had them calculate with it. Why? Because the theory has become far from worthless. You can see the reason by comparing Figure 15.5 with Figure 15.3. The red rectangle in Figure 15.3 is reflected in the later image in a warped version. Perhaps you can imagine: as long as you perform calculations only within the red rectangle, no one notices the difference, in the same way that an architect designing a house does not have to take into account the curvature of the ground because Planet Earth is spherical and not flat. On a small scale, Newton's theory is exactly as good as Einstein's, but much easier to work with, because the necessary mathematics can be mastered by students at school, and that is certainly not the case for the mathematics of curved spaces required for 'Einstein'. Newton is definitely not obsolete; his theory still works fine.

As an aside, the image I used as a source for figures like Figures 15.3 and 15.5 is a *fractal*, a mathematical figure about which you can find more in Appendix 25. You can download the computer program I used to make them, so you can make such figures yourself.

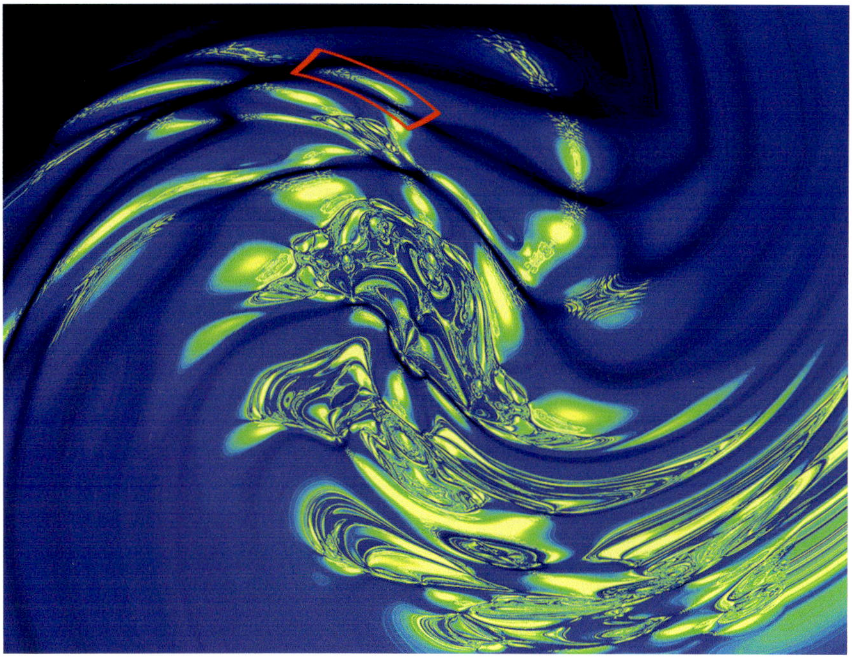

FIGURE 15.5 After a scientific revolution.

The conclusion is: if successful scientific theories are overtaken by new theories, whether by revolution or not, they retain their value: the change is not that they are discarded but that a limitation is imposed on the area in which they can be applied. This makes sense, since they have been extensively tested within that limited area and have proven their worth there.

But if protoscientific models are overthrown in a revolution, they are subsequently abolished and forgotten – unless they persist in a shadowy realm of pseudoscience and superstition, in accordance with the *Law of Conservation of Belief*. Think of: astrology, iridology, homeopathy, creationism, voodoo, conspiracy theories.

Naturally, the logical next question is: what does all this mean for the protoscience of economics? What physical and mental tools are ready to bring about the revolution that will bring about the breakthrough to science? What words will be used; what stories will be told? Above all, if we succeed in developing a mature omniconomic science, what will happen to current economic models? With the policies of governments and corporations? With humanity, nature and the planet?

Peaceful

Before I address those questions in the following chapters, I emphasize here in advance that where I write about revolutions, I am referring *exclusively* to scientific or technological revolutions and not to violent social upheavals. Both the French and Russian Revolutions caused great misery, and I know of no example of a disruptive revolution in society that did not. I'll come back to this in detail in the final chapter; but I'll tell you in advance that the goal of this book is to contribute to a revolution in economics as a field of study and thus to a healthy, peaceful evolution of human society.

Recommended further reading (for details, consult the online Bibliography)

Journal article: Morgan, Mary S. (2001): Models, Stories and the Economic World.
Book: Kuhn, Thomas S. (1962): *The Structure of Scientific Revolutions.*
Book: Lakatos, Imre (1978): *The Methodology of Scientific Research Programmes.*
Book: Feyerabend, Paul K. (2004): *Problems of Empiricism.*

16

CHARACTERISTICS OF OMNICONOMIC SCIENCE

Now first a brief recap. It is good to list some of the necessary features and prerequisites of the proposed omniconomic science, as they have emerged in the previous chapters. I will refer many times to earlier page numbers, so that you can easily look them up if you wish. In a few cases, I will also refer ahead, to subsequent chapters.

In surveying the diverse characteristics and conditions for the omniconomic science to be designed, it appears that they can be divided into four groups, each with four criteria, one for each sphere.

Characteristics of a successful omniconomic science
P = characteristic of Protoscience (~~undesirable~~); S = characteristic of Successful science (desirable)

	Scientific *Mature* *(beyond proto)*	**Systemic** Intellectual output	**Sustainable** Societal output	**Engaged** Emotional output
Phenosphere *Effects*	***Empirical*** ~~P6. Lack of empirics~~ ~~P7. Lack of success~~	***Unifying*** Tegular S3. Theoretical edifice	***Peaceful*** Revolutionary S6. Contributions to society	***Accountable*** to society S9. New questions
Politosphere *Policies*	***Methodical*** ~~P4. Copied methods~~ ~~P5. Oversimplification~~	***Multilevel*** Interaction between levels	***Strategic*** Route to target S5. Technology	***Purposeful*** Participatory democracy
Mythosphere *Narratives*	***Reflective*** Not eagercredulous; Critical judgment ~~P3. Myths~~	***Complete*** Closed system S2. Understanding	***Dynamic*** Broad Holloway S4. Predictive power	***Inclusive*** Transdisciplinary S8. Insight into who we are
Logosphere *Words*	***Coherent*** Common canon ~~P1. Struggling schools~~ ~~P2. Impetus words~~	***Complex*** Emergence, attractors S1. Order	***Intrinsic*** Not glued-on	***Values-based*** Intersubjective S7. Amazement, beauty

DOI: 10.4324/9781003306269-18

With this, I distinguish a total of sixteen criteria by means of which the degree of success of a (hopefully soon) starting and growing new science can be assessed. The set of sixteen criteria can be used as an assessment which can also be applied to existing sciences, as I show in Appendix 26. I discuss the four groups of criteria one by one.

16.1 Scientific criteria: coherent, reflective, methodical, empirical

The first group relates to scientific criteria. In short, these come down to the fact that the protoscientific stage has been passed. That is why the seven characteristics of protoscience can be found in the chart of characteristics, divided over the four spheres in the way as described earlier at the beginning of Chapter 11. They are shown here in a strikethrough font, by which I symbolize that science should not possess these seven characteristics at all.

The characteristics in this column of the diagram apply to any science, not specifically to omniconomics. I go through them briefly one by one, from bottom to top.

Coherent

A first characteristic that distinguishes any mature science from a protoscience is that there is a core area that is widely accepted scientifically: the shared edifice (page 91), the canon (page 212). There is no struggle over essential aspects; there is not a range of schools as in 19th-century biology (page 60) and 21st-century economics (page 61).

Such a canon is possible because its various parts are mutually consistent, so that the discipline is coherent, and thus contains no inner contradictions. Terms in the canon that might once have had an impetus character have been deleted or fundamentally modified (page 46).

Reflective

Sound science is not based on *eagercredulity* (page 73) or prejudice (*bias*, page 197): every hypothesis is critically tested, not once but continuously. *Falsifiability* (page 197) is the touchstone; but the falsifiability requirement is not taken too strictly, since then creative speculation would have no chance (page 216).

Methodical

Every science, including omniconomics, employs its own appropriate methods, not directly and uncritically copied from other fields of science (page 80). Mathematics is often a useful tool, but by no means in the same way everywhere (page 80). The same is true of observational methods and tools (page 208).

The theoretical constructs are set up in a well-structured manner. The explanatory principles applied (page 196) are explicitly stated.

Empirical

The empirical testing of each hypothesis and theory is central. The testing is done through experiments and observations (page 191). *Replicability* is an absolute requirement (page 191).

From the observations, by means of induction (page 175), hypotheses and theories are constructed (page 194), which leads to insights, thus to *understanding*, but never to eternal or absolute truths (page 176).

16.2 Systemic criteria: complex, complete, multilevel, unifying

In the other three groups, consisting of systemic characteristics, sustainability and engagement, you will find the success factors introduced in Chapter 8, shown in green.

To begin with, here are the systemic criteria, which relate to the *intellectual output* of science. As an aside, we are not talking here about 'systematic' (in the sense of ordered, structured) but about 'systemic', that is, about science characteristics related to the system that is being scientifically investigated.

Complex

The research field of omniconomics comprises very many elements, varying in size from minute electrons to entire planets and the Sun, and thus together they span a large number of aggregation levels. Within them is all of earthly life, all of humanity, the Internet, and all rivers, roads, and power plants. Together, these form a churning, dynamically swirling world. The omniconomic system is by definition complex and should be treated as such, without oversimplification (page 83). Fortunately, we now possess the necessary tools to do so: 'complexity' is the organizing principle that enables us to recognize meaningful patterns in such systems, using new words such as 'emergence' and 'attractors' (Chapter 17).

Complete

Today's economics employs *externalities* (page 34), as a result of which the economic system under study 'leaks': apparently free goods enter the system from nature, while all kinds of waste leave the economic system, also for free to all appearances. The 'leaking' economic system is not a *closed system*, that is a system where it can be reasonably determined that no meaningful things – materials, energy or information – move in or out through the boundaries of the system. If it is not possible to describe a system without using externalities, that system is incomplete and must be made larger, until everything that is relevant is part of it and the system is thus complete.

Multilevel

Just as the system studied must be complete in a physical, spatial sense, so it must be in the 'fifth dimension', that of scales. Omniconomics is complete, only if it involves all relevant levels of aggregation. Thus, the *actors*, the 'players' in the system, are not just individual people (page 202), institutions, or countries (page 63), but also ecosystems, the Moon, the planet Mars, and the Sun; nanorobots (in development), bacteria, viruses, and even individual atoms and electrons. In the *Universal Atlas of Scales* (Appendix 24), relevant aggregation levels range at least from −6 to +4, with sizes ranging from 10^{18} meters to 10^{12} meters: a *petameter*, a trillion meters.

Unifying

Omniconomics must be unambiguous in drawing conclusions and making recommendations regarding concrete policies.

Physicists have been striving for *unification* in the form of one comprehensive *Theory of Everything* for at least a hundred years (page 216). However, the omniconomic field of research is so complex, and so diverse in levels of aggregation, that unification can only be obtained through a combination of various theories and models that complement each other from different levels of aggregation, viewpoints, and explanatory principles. You can compare this to the way a roof is covered watertight, not by one large entity but by a collection of well-matched and partly overlapping roof tiles. I speak of *tegular* (tile-like) *unification* (Chapter 18).

16.3 Sustainability criteria: intrinsic, dynamic, strategic, peaceful

The characteristics related to sustainability deal with a different kind of scientific success: *societal output*.

Intrinsic

The main goal of the omniconomics is an *intrinsically* sustainable society (pages 33, 161) in which sustainability is not glued on but an inseparable, integrated aspect of society: a feature of our identity as a community.

Dynamic

Omniconomics does not lead to a static world (page 37) in which, because of an achieved safe balance, nothing may ever change again, but to *dynamic stability* (page

40), represented in the image of the *Broad Holloway*. This does not guarantee that society can never collapse, but minimizes the likelihood of this via disastrous tipping points (page 138). Conditions for this include an end to continued economic growth, or at least material growth (page 99), and an ecological footprint no larger than the biocapacity (page 100).

Strategic

Designing an intrinsically sustainable society, harmoniously embedded in nature, is one thing. But omniconomic science has a second task, which is to discover and realize a pathway to achieve that society. There is every chance that this 'task 2', finding and following that route, will be at least as difficult as 'task 1', defining the end point of the route. Needless to say, you can expect all kinds of resistance to emerge, especially from those who feel – rightly or wrongly – that they benefit from the current established order.

One of the tasks of the omniconomic science will be to find a way to let as many people and institutions as possible gain from the transition, and also to give them the *feeling* that they are gaining from it (Chapter 20). It goes without saying that the current impetus word 'profit' will have to be defined much more broadly than just financially (page 101).

Peaceful

This brings up another precondition, which I mentioned on page 219. Scientifically, omniconomics will undoubtedly be *disruptive*: a tearing apart of the present. Societally, however, the transition should definitely not be disruptive, since that would lead to indescribable destruction and misery. The route to intrinsic sustainability should not be a revolution but an evolution. It belongs to 'task 2' of omniconomics to discover how a scientific revolution can societally turn out to be a healthy, peaceful evolution of human society towards intrinsic sustainability.

16.4 Engagement criteria: values-based, inclusive, purposeful, accountable

The final group of characteristics of omniconomic success, the engagement criteria, relates to the *emotional output*. This output will be discussed in Chapter 20; here is a preview to complete the list of characteristics.

Values-based

Omniconomics is not a neutral branch of science. Like medical science, for example, omniconomics strives for goals formulated by or from society. Omniconomic science operates from involvement with society and is therefore by definition values-based (page 200) and action-oriented, built from a societal mission. This values base is to be explicitly accepted and cherished.

Since omniconomics is partly about values such as beauty, passion, and love, which cannot be judged objectively, there is a key role for *intersubjectivity* and consensus building.

Inclusive

Since omniconomics aims at the realization of societal goals, the right to set those goals belongs to society: the world belongs to all of us together. This means, that *everyone* possesses the right to participate in that determination: the process is inclusive, no one is excluded. Omniconomic science is therefore not only an interdisciplinary process: laypersons, groups and organizations, not related to any scientific discipline, also participate actively in the research. Such science is called *transdisciplinary.*

This inclusive collaboration between theory and practice will strengthen the self-awareness and identity of individuals, groups and peoples, and thereby contributes to the image of *who we are.*

Purposeful

According to the Four Spheres model (page 21), *words* determine *narratives*, which in turn evoke *policies* that produce *effects* throughout society: see Figures 3.3 and 3.4, among others.

It is these *effects* that guide omniconomics as a science. After society reaches conclusions about the design of society through processes of *participatory democracy*, it is the task of omniconomics to reason back through the four spheres: through desired effects to policies, to narratives, and to words. That goal-directed process will, of course, not only proceed in one direction, since omniconomic research will subsequently reveal the extent to which the effects desired by society are attainable and at what rate. All in all, there will be an iterative, never-ending process in which the desires, ideas, and possibilities will undulate back and forth between society and the researchers.

Accountable

Omniconomics is owned by society. That means that researchers are accountable to that society, and this requires transparency. Only ethically sound research methods are used (page 87), in the form of independent, publicly funded research. As a rule of thumb (page 199): ignore all conclusions of research paid for by stakeholders. And rule out any form of pseudoscience (pp. 60, 107).

And now, let's go

After this inventory, it's time to look at what new tools are at our disposal, and what we can accomplish with them. That's a lot.

Recommended further reading (for details, consult the online Bibliography)

Journal article: Nagatsu, Michiru et al. (2020): Philosophy of Science for Sustainability Science.

17

THE COMPLEXITY OF THE OMNICONOMIC WORLD

From this point forward, I am going to deploy a series of new tools for the purpose of forming omniconomic science. To begin with, that involves an impressive collection of *physical* tools (= step 1 on the *Path to Science*) of an electronic nature, such as computers, Wi-Fi, satellites, the Internet, neural networks, massive data storage and processing.

Thanks to them, we have at our disposal a gigantic array of *mental* tools (= step 2), and it is these tools that I apply in the coming chapters: complexity, simulations, network theory, artificial intelligence, self-learning systems, big data, virtual reality, artificial life. With their help, we now have at our disposal new patterns, new words, and wonderful new narratives.

17.1 New equipment: complexity

One of the characteristics of omniconomic science mentioned in Chapter 16 is that it is realistic, that is, it represents reality well, without oversimplification, on the premise that the omniconomic world is a complex system. This is a fact with far-reaching implications.

Five types of systems: from static to chaotic

Systems can be classified into five categories, by paying attention to the degree to which something *happens* in the system. The simplest category is that of *static* systems, which are rather boring since nothing happens at all. Think of a brick in a warehouse: static systems are very predictable.

A little more interesting are *stationary* systems. In these, activity does occur, but if you wait a while, you will discover that this activity goes around in circles and

DOI: 10.4324/9781003306269-19

returns to its original state a little later. Look at a clock for a few days. Or look at a stationary car whose engine is idling for fifteen minutes, and you will see exactly what I mean. The predictability is still high: it may be nice to observe the progression for a while, but in the long run you will get bored again.

The third category is that of *dynamic* systems. These are more interesting, because there is actual development, although it is often still fairly predictable. For example, as I reread the previous paragraph after typing, I have a glass in my left hand that I hold one meter above the ground. If I let go of that glass, I can calculate (using Newton's laws), very precisely, at what time the glass will hit the ground. You can also look at the already mentioned car that is now calmly driving across a parking lot, causing a real displacement. There, too, you can make a reasonable estimate of when the terrain has been completely crossed.

But now for the fourth category: the *complex* systems. As soon as the car leaves the parking lot and enters the busy traffic, it enters an extremely complex interplay of many 'players' and forces, such as other cars, bicycles, traffic lights and road signs, roads and curves, potholes and speed bumps, trams, sun or rain, billboards and road detours. Now Newton's laws are of little use to me; they don't help me predict which object will be where and when and whether everything will end well.

Complex systems are so complicated because the various players and forces interact back and forth.[323] This can have a self-reinforcing effect, for example when motorist A reacts in an irritated manner to moped rider B, who points to his forehead, which makes A even angrier and more aggressive, so that B is forced to kick a dent in a car door. Self-reinforcing effects – positive feedbacks – can easily get out of hand and lead to undesirable situations.

You could see another example of this in Figure 9.2 (p. 119), where, as a result of positive feedbacks, the poor get poorer and the rich get richer, so that inequality naturally tends to increase. Climate change is an even more complicated example, as Figure 17.1 shows, in which many of the feedback loops drawn reinforce each other's effects, while some, thanks to negative feedbacks, actually moderate them. The combination of all these causes and effects is that climate change is very difficult to predict.

And then, finally, there is the fifth category, that of *chaotic* systems. In these, the parts of the system exert such forces and interactions on each other that the system behaves in a completely unpredictable way.

Such chaos can occur suddenly, for example at the moment when the glass I dropped hits the ground and splinters into seven thousand shards and splinters. There is no way to predict in advance what the pattern of glass splinters will look like after impact. Perhaps you would expect that, given enough computing power, a supercomputer could calculate the details of the disaster area in advance if I provided the computer with accurate information about material, shape and size of the glass, the position in which I released it and so on. That is not true, as it turns out when I let such a computer do the calculation twice: the second time with a tiny, imperceptibly small difference from the first time, for example with a deviation in

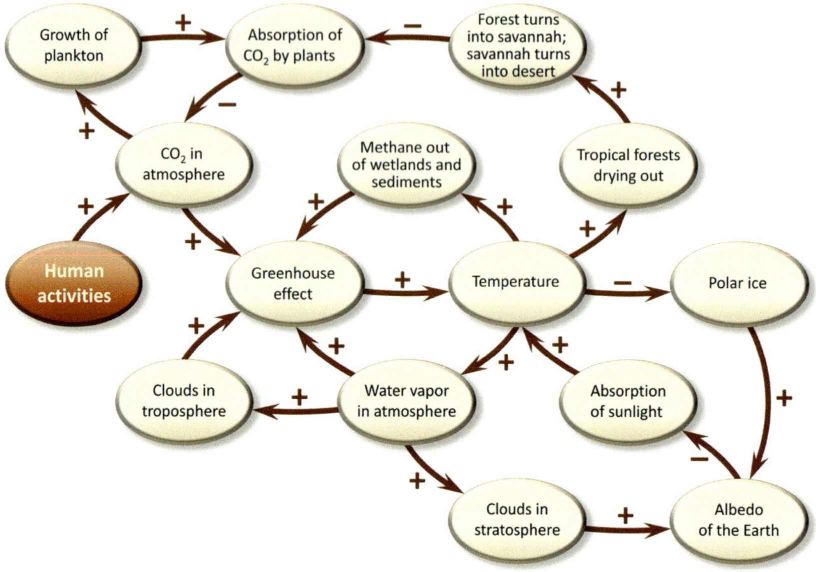

FIGURE 17.1 Feedback loops related to climate change.[324] (albedo = reflective power of the planet)

the angle in which I held the glass of less than 0.001° or with a temperature difference of 0.001°Celsius or Fahrenheit in the glass which makes the material properties slightly different. The calculated end pattern of the shards the second time is not a little, but *completely* different from the first.

In a chaotic system, imperceptibly small deviations in the situation of a certain moment can grow into gigantic differences in the situation some time later. A famous example of that principle is the 'butterfly effect': the anecdote of the butterfly in the Andes that causes a hurricane in Western Europe with one stroke of its wings. That story is not really true, because the butterfly does not *cause* the hurricane. But the difference in air pressure that the butterfly's wings evoke is a link in a branching chain of atmospheric processes that proceed slightly differently than without that wingbeat, after which that deviation grows and grows in the course of days and weeks until the development is completely different, which in a completely different part of the world can mean the difference between beach weather or houses blown away. The global atmosphere, and thus the erratic weather, is one big chaotic system in which detailed forecasts are very difficult to make, as anyone who has ever been caught unawares by a downpour knows.

Chaos theory, which describes this kind of behavior, is a relatively young scientific model that has proved successful in numerous disciplines – not so much in making predictions, but in understanding phenomena, including unpredictability – thus bridging those fields of science using common *words* and *narratives*.[325]

Emergence on the edge of chaos

Of the five types of systems, the complex systems are the most interesting. And of those complex systems, the most exciting are those that are almost chaotic, that is, those that are near the *edge of chaos*.[326] Such fascinating systems are what you get to see when huge numbers of 'players' (*actors*) interact in an extremely complex way. The result is a collection of entirely new, surprising properties that *emerge*, seemingly out of nowhere.[327]

Emergence of new properties occurs even in much less complex systems, provided enough actors are brought together. An accumulation of 10^{25} molecules of H_2O in a cup is *wet*, while an individual water molecule is not wet. 'Wetness' is simply not a property of molecules; it is a property that emerges spontaneously in certain systems of very many molecules together.

Emergence is only really stunning, when lots of actors act together in a system so complex that it behaves almost (but just not really) chaotically. It then seems as if an entirely new kind of matter magically or supernaturally emerges. Put a hundred billion (10^{11}) carefully selected molecules in a balanced way into a microscopic skin, and you have a *living bacterium*. If you expertly assemble a thousand times that many molecules, you even have a real living body cell. Or a wandering amoeba. 'Life' comes across to many people as so magical or mythical that living matter has been seen through the ages as a separate kind of matter, filled with *odyle*, *vis essentialis*, *Bildungstrieb*, *élan vital* or some other kind of divine power, as Chapter 5 showed (p. 47).

Go one step further. Now glue together about ten or a hundred trillion (10^{13} to 10^{14}) of those living body cells in a thoughtful way, and you get an animal, which, according to the Greek philosopher Aristotle, possesses a *sensitive soul*, capable of sensations and feelings.

It can be even more fantastic. Take, when building such an animal from living cells, among other things, one hundred billion (10^{11}) brain cells, together accounting for just under one and a half kilograms (or three pounds). Connect them in a clever way via some hundred times as many signal wires, and you get a human being, a *homo sapiens*, who, according to many, is in possession of an *immortal soul* breathed into them by a God. This is no wonder, because humans (and quite a few other animals[328]) again possess new emergent characteristics; they can be kind, for example, or intelligent, or nervous. They possess self-awareness and free will – although the latter is disputed. Their constituent cells individually have none of that, let alone the molecules that make up those cells: have you ever heard of a liver cell in love or a poetic protein molecule?

It's like a fairy tale, but really true: emergence in complex systems can have mind-boggling, wonderful effects.

What you see here is always: the transition from a lower to a higher aggregation level, with new properties emerging each time that are characteristic of that level. In the *Universal Atlas of Scales* (Appendix 24), in which each level is a thousand times as large as the previous one in terms of size, molecules are at level −3, bacteria at

−2, body cells more or less between −2 and −1, humans at 0. On this last level, that of us and our fellow animals, phenomena such as friendship, love, caring, poetry, music and dance have emerged over millions of years: in humans as well as in several other species.[329]

It did not stop there for, in turn, cooperating humans created wonderful new complex systems at higher aggregation levels: village communities, educational institutions and businesses at level 1; countries at level 2. In doing so, new unexpected properties emerged, including culture, science, and solidarity. Using our technology, we then built even larger complex systems, including continent-wide road systems and electricity grids. We connected ten billion (10^{10}) computers and other devices via signal wires and Wi-Fi, creating the Internet. And we created the international economy.

Humans were not the only ones to build such vast complex systems; nature did so long before we existed. A pond and a forest are complex systems, as is a food chain. At a higher level of aggregation, the global biosphere is a complex system, which includes humans and everything humans created.

17.2 New words: homeostasis, attractors

The closer a complex system is to the edge of chaos, the more mobile and exciting its developments and emergent properties are. This is wonderful, but there is a downside: it doesn't take very much for such a system to tumble over the edge into the realm of chaos.

If that happens to an individual human body, it can become seriously ill or even die. If it happens to a person's brain, a momentary epileptic seizure – a wild electrical storm in one or more brain regions – or a psychotic state can break out.

In a road network, an accumulation of queues of cars can lead to a traffic infarct. A computer, also a complex system, can crash hopelessly; have you met the infamous 'blue screen of death'? Parts of the Internet can be jammed by a DDoS attack, a society can erupt into a frenzy like Rwanda in 1994, or into a bloody revolution like the French in 1789. The global community may even descend into a world war.

Tumbling over the edge of chaos can be slow and insidious, or, on the contrary, abrupt through a tipping point (which presumably was preceded by a creeping process). That tipping point may be the collapse of an ecosystem, caused or followed by a massive extinction wave of plant and animal species. It could also be a global economic crash in which stock prices plummet and banks, companies and countries go bankrupt.

Homeostasis

Life, humans, human society, the Internet, ecosystems and the biosphere: all these systems are enchantingly beautiful and at the same time vulnerable, because all of them are near edges of chaos. Systems that aren't, are safer but much less interesting. But fortunately that is not the whole story, because all these systems are – to some extent – able to protect themselves from impending chaos and to repair themselves.

Humans and other animals defend themselves against invasions of germs with their immune systems. An additional tool for this is fever, which the body deploys to kill invaders. Once that is successful, the body turns the internal thermostat back to the natural equilibrium temperature of 37°C (98.6°F) and the body returns to *homeostasis*, a state that can be stably maintained because the various physical and chemical processes keep each other in balance.

Thanks to the narrative called 'complexity', we can apply the word 'homeostasis' to other complex systems, where the term was not always used before. You recognize it in the 'natural balance' and resilience of forests, seas and other ecosystems. Homeostatic balance is present in ants' nests that respond flexibly to changing conditions.[330] In individuals who resiliently adapt their lives when their jobs, families or health change dramatically. In businesses that are adaptable enough to do the same, and in democratic societies in which a careful separation of powers – government, parliament, judiciary, free press, empowered civil society – ensures the continuity of freedom and rule of law in a changing world.

Homeostasis is a dynamic equilibrium, in which a system responds flexibly to change and thereby manages to maintain its essential properties, like a tree that bends and therefore doesn't break in the wind.

Attractors

The body temperature of 37°C in healthy mammals is a kind of *attractor*, an equilibrium temperature to which the body keeps returning 'as if attracted towards it'. Such an attractor is not really there, because the temperature itself does not 'attract': it is the result of numerous forces and processes that, in complex, negative feedbacks, lead to equilibrium. The word 'attractor' is a metaphor.

Numerous systems possess such attractors. For the glass I released, the ground is an attractor (although the actual attractor is the entire Planet Earth, which literally attracts the glass with its gravity). For a hungry person the kitchen is an attractor; for a lover the beloved; for a compass needle the North Pole; for a junkie the hypodermic needle.

Not all attractors are stationary equilibrium states. Homeostasis in a human body exhibits variations over the 24 hours of a day and night, with metabolism, temperature, and brain activity forming a cycle: the daily biorhythm. Natural balance is also a cycle, with the annual rhythm of the seasons. If a resilient system gets out of an equilibrium cycle because of a disturbance, it will slowly but surely recover that cycle – unless the disturbance is too great and something breaks down permanently.

You can think of attractors as the *tendencies* of a system: that which a system attracts to if you let it run its course. *Equilibrium attractors* lead to a healthy and dynamic equilibrium. But some attractors actually tend to destroy the system. For my falling glass, the ground is such a *devastating attractor*, as is the hypodermic needle for the junk.

Devastating economic attractors

When you let the economy take its course, the system by itself tends towards inequality: wealth makes richer and poverty makes poorer. *Inequality* is an attractor that threatens the social equilibrium. This is a fundamental system error, which stems from the principle 'money attracts money', as Chapter 9 showed, depicted in Figure 9.3 (page 120) in the form of snapshots of a movie.

Wealth makes one not only richer but also more powerful: concentrating *power* is a dangerous attractor that leads to manipulation, tyranny and war. Unfortunately, this is not theory but reality, as historical and current developments amply prove.

The stock market also possesses devastating attractors. Rising stock prices increase the confidence of traders, causing stock prices to rise. This self-reinforcing effect causes stock market *bubbles* that are risky attractors. The opposite is also true: falling stock prices cause confidence in the stock market to decline, so the decline is self-reinforcing too. Consequently, economic *crashes* are also attractors of the economic system, which is therefore fundamentally unstable.

Economic growth reinforces itself because growth creates capital to invest in growth. Continuous growth is even a necessity in the current economic system to prevent economic collapse, as I described in Chapter 8 (page 96). Growth is a devastating attractor, too.

As a result of these and other devastating attractors, the macroeconomic system is about as unstable as a drinking glass on its way to the ground: it is guaranteed to break, unless you intervene.

Complexity theory, then, offers here a brilliant example of how new words make it possible to look at reality and make plans in an entirely new way. That's what enables me to write:

> The instability of the global economic system is caused by *devastating attractors*.
>
> Therefore, the omniconomic mission is this:
>
> To create intrinsic sustainability, by redesigning the global human system in such a way that these destructive attractors give way to *equilibrium attractors*, with the effect of creating a globally resilient *homeostasis*.

Is that possible? I think it is, and in the rest of the book I want to show it. I begin with an outline of research on attractors. Such research is done with another new tool: computer simulations.

17.3 New equipment: simulations

In all branches of science, simulations have become indispensable tools. In astronomy, the Big Bang and the formation of galaxies, suns and planets are simulated on the very largest supercomputers, which often spend days and nights calculating.

Meteorologists use simulations to calculate weather forecasts. In Chapter 15 (page 209) I mentioned the example of a medicine, a painkiller, of which the effect and the side effects are calculated in the computer. I am not exaggerating when I say that simulations have become indispensable to any science. The film about 'money attracts money' was also made with a computer simulation that I wrote for that purpose, which you can download to create more of those simulations yourself.

In general: without computer simulations, most sciences would be nowhere near where they are today.

Simulations have been used to do a lot of research on attractors in complex systems. One of the oldest examples is the cycle of *predators* and their *prey*. I will say right away: the model I am about to describe is instructive, but it is definitely an oversimplification of reality. I will come back to that later. Details of this and the following computer models are available in Appendix 27.

Imagine a simple ecosystem in which only foxes and rabbits live. The rabbits live off the grass that is widely available. When their numbers increase, it is pleasing to the foxes who consume them, so a little later their numbers increase as well, as the red curve in Figure 17.2 shows. The peak of the fox population comes a while after that of the rabbits. Around the time that the number of foxes peaks, so many rabbits are devoured that their numbers decline rapidly, as a result of which many foxes die of starvation. This causes the rabbits to crawl out of their crisis and grow in numbers again, so history repeats itself and a cycle is created.

You can also depict the process in a *phase diagram*, as in Figure 17.3, in which the numbers of predators are placed on one axis, and their prey on the other. In that diagram, the cycle is run counterclockwise, as indicated by the arrows.

In certain theoretical models of predators and their prey, the situation is so stable that the cycle acts as an equilibrium attractor to which the system spontaneously returns after a disturbance. The attractor is then called a *limit cycle*. I illustrate this with a few examples.

To begin with, there is Figure 17.4, in which some catastrophic cause – forest fire perhaps? – in a short time simultaneously causes the numbers of foxes and rabbits to drop sharply, as shown by the red arrow. As a result, the state of the system suddenly ends up somewhere inside the limit cycle. In the period that follows, you

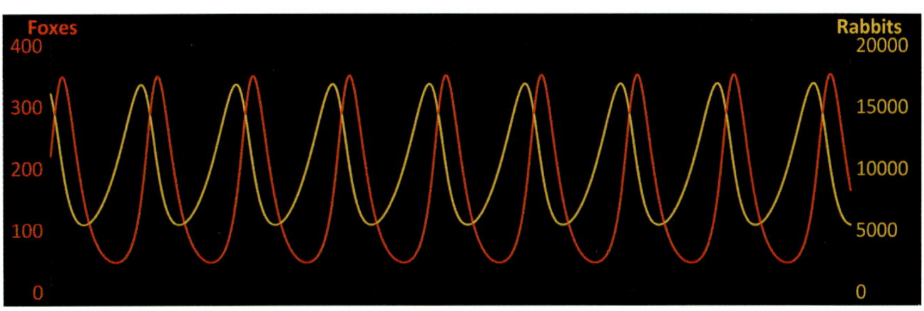

FIGURE 17.2 Foxes and rabbits: a cycle.

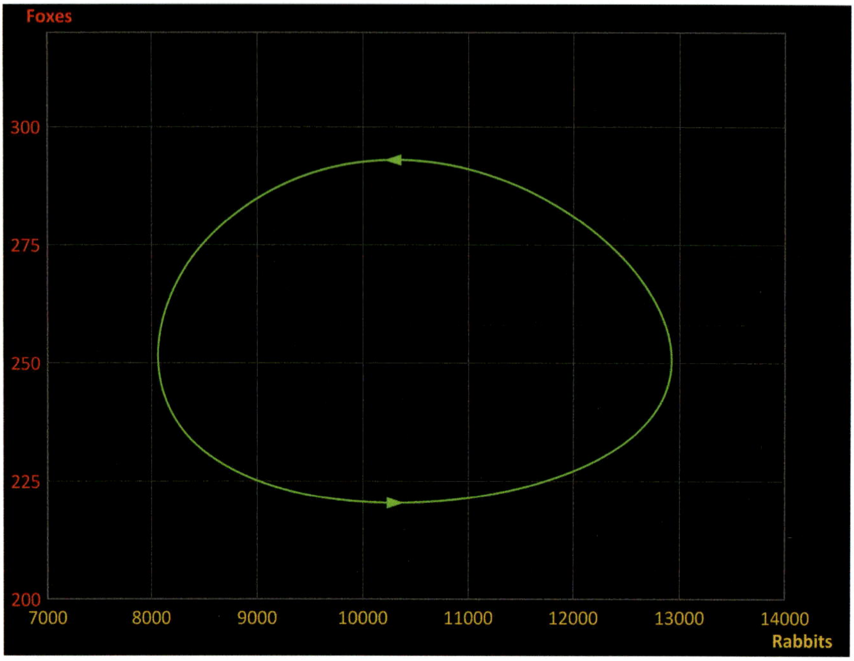

FIGURE 17.3 Foxes and rabbits: phase diagram of a limit cycle.

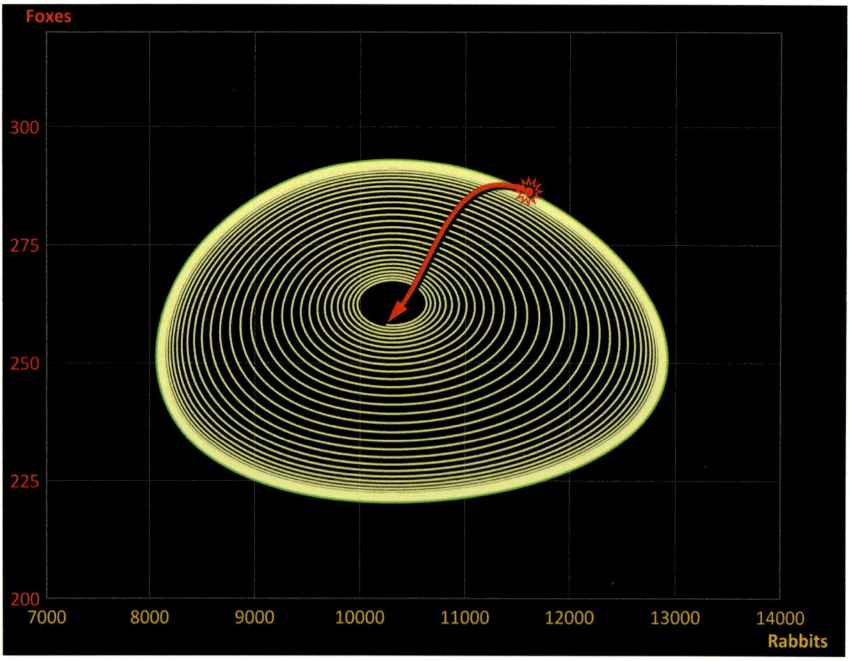

FIGURE 17.4 Foxes and rabbits: first disturbance.

can see how – according to the simulation model – the cycle slowly increases in size again and eventually returns to the limit cycle via a spiral, which therefore indeed turns out to act as an equilibrium attractor that is recovered by the system.

This stabilizing function also appears after another disturbance, shown in Figure 17.5. In it, the number of rabbits declines in a short time due to an epidemic, while foxes are not initially harmed. But next they are, due to lack of food. This time, the state gets well outside the limit cycle (which is now shown much smaller because of a different scale of the graph) but again resiliently returns to the equilibrium cycle. Thus, the system appears to be considerably stable.

The two figures illustrate: the system is in a dynamic equilibrium, in homeostasis: it is able to recover from disturbances. This makes the system of the foxes and rabbits a nice metaphor of the intrinsically sustainable society that is the goal of omniconomics. Please remember: it is an oversimplified metaphor, proving nothing but only illustrating something.

However, despite the dynamic stability, disturbances are also possible that definitively destroy something. This is evident after another disaster, where the habitat available in the computer model is halved, perhaps because half of the area is cleared for the purpose of building a highway or a residential area. In the process, all foxes and rabbits are chased away to the remaining half of the nature reserve, so that their numbers initially remain the same. Unfortunately, as the computer model

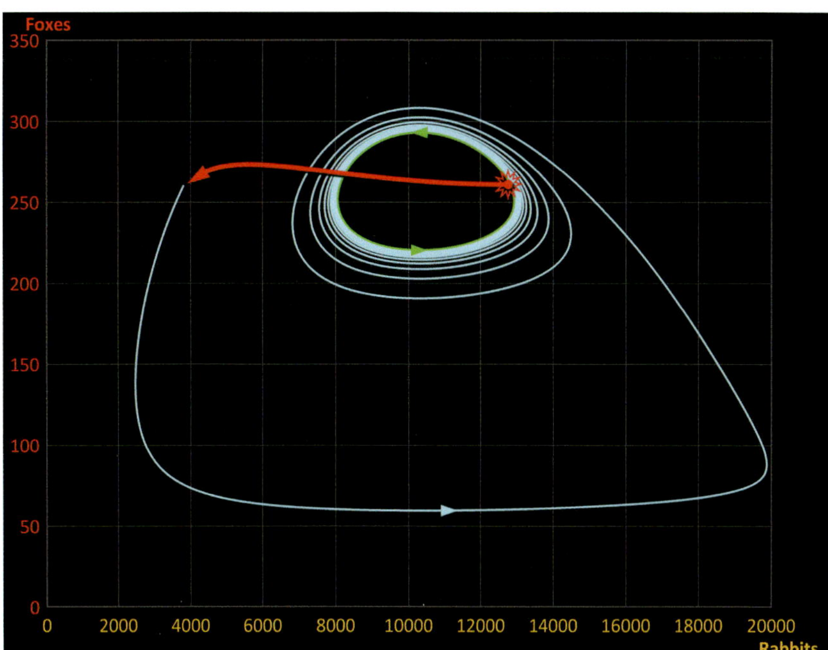

FIGURE 17.5 Foxes and rabbits: second disturbance.

calculates, this time the system fails to return to a stable cycle, as Figure 17.6 shows. From the initially stable limit circle shown in green, the purple line from the tipping point (the asterisk in the figure) draws the disastrous outcome. It turns out that it is the foxes that are most affected by the transition: they die out completely. The end point of their route, represented in the figure by a small dot at the bottom, now appears to act as an attractor, but of the devastating type. That this is indeed a genuine attractor becomes clear when the simulation is repeated from other starting points. Each starting point appears to draw towards the same devastated state in which the foxes have died out.

You can experiment with this simulation model yourself, because the images shown were created with a computer program (*Fox Rabbit Math Model.exe*, a Windows program) that I wrote for that purpose. You can download it from the website that accompanies this book. The program allows for many more variations, including randomness, as Appendix 27 explains.

Tipping points

The last, catastrophic disturbance of the fox and rabbit system illustrates another important phenomenon: the swing from an old to a new attractor via a tipping point.

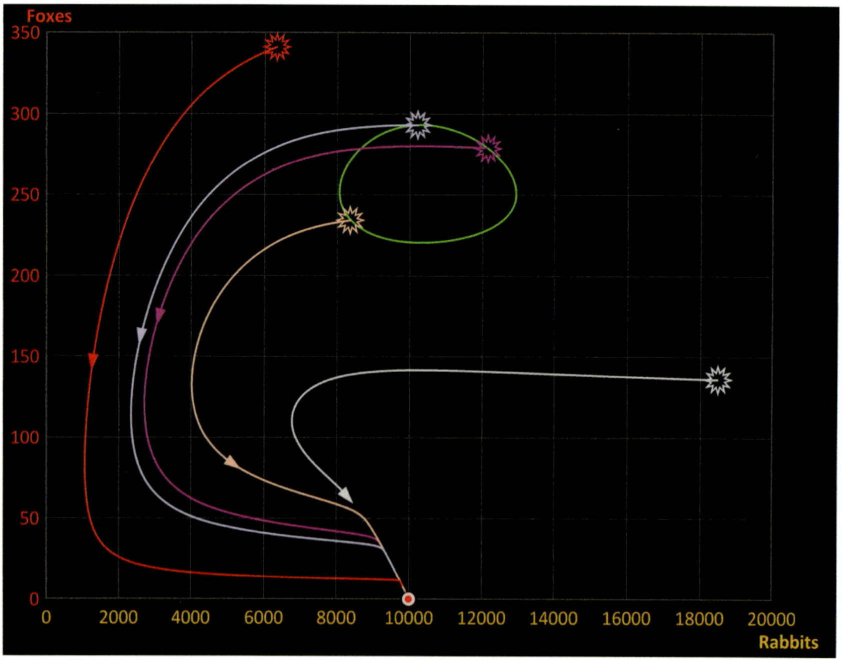

FIGURE 17.6 Foxes and rabbits: third, devastating disturbance.

An example of this in the real world is a possible change in climate from a *cold* stable climate cycle to a *warm* one. It has been known for quite a long time that Planet Earth has experienced a number of ice ages over the past millions of years during which the average temperature was about six to eight degrees Celsius lower than it is today. In the past 740,000 years alone, there have been eight such alternations of ice-cold and temperate periods.[331] It seems likely that this has involved repeated cycles that have more or less the character of a limit cycle, that is, of a stable attractor, with us and our planet currently near the warmest point of the present cycle, as Figure 17.7 shows.

Human intervention is currently causing a further continuing rise in temperature. Research has been done into the possible consequences,[332] and they could include the cycle reversing to a new cycle where the current temperature is not the *warmest* point but the *coldest* – as if by passing a tipping point a switch is flipped. This (as yet happily unproven) hypothesis is called *Hothouse Earth.*

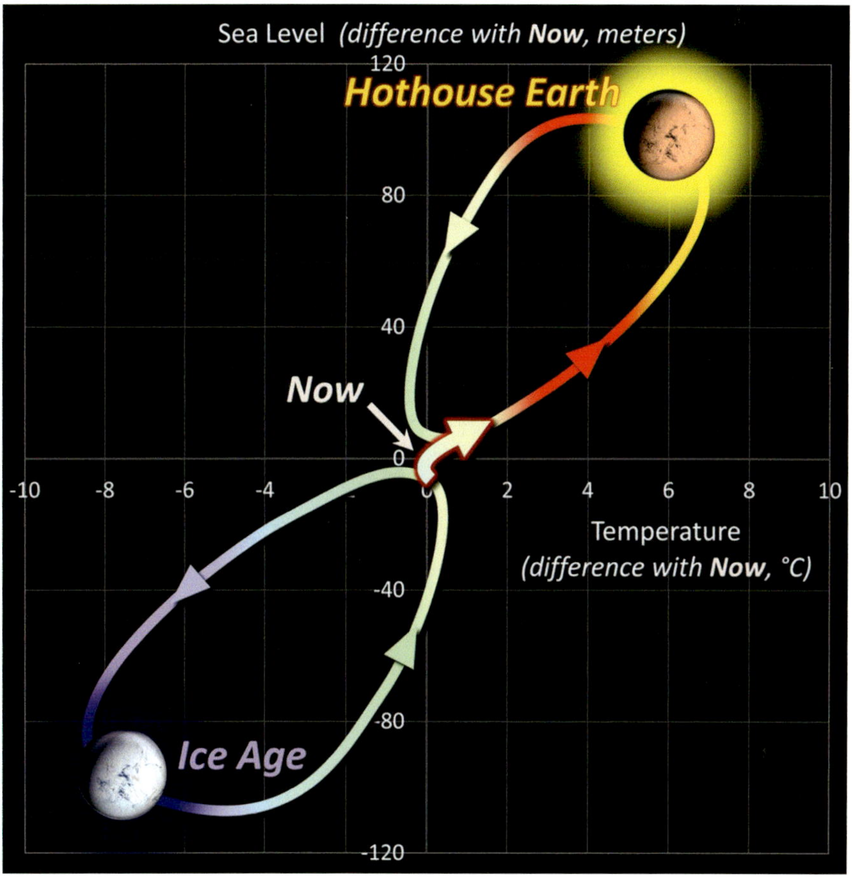

FIGURE 17.7 Ice Ages and Hothouse Earth: the phase diagram.

If the Hothouse hypothesis is correct, the now inevitable temperature increase of at least 1.5 to 2°C will not then move back, but irresistibly continue to rise to 6 or 8°C, which would presumably totally destroy human civilization.

Model and reality

I wrote it already: the model of the foxes and rabbits is a greatly simplified representation of reality. While its diagrams are interesting and instructive, a habitat in the real world is many times more complex. It has been investigated whether limit cycles really exist in such a context, and there is some evidence for that, for example with respect to lynx and hares in Canada. I discuss this in more detail in Appendix 27.

More important is the question: to what extent does the Hothouse hypothesis correspond to reality? Here, too, there appears to be a considerable oversimplification. To begin with, Figure 17.8 shows for the last four ice ages that the average temperature did not go smoothly up and down in a cycle of 100,000 years. The course was much more erratic.[333] Then, if you combine that temperature trend with data on sea level, you can compare the tight egg-shaped curve in the Hothouse diagram of Figure 17.7 with the actual graph in Figure 17.9, based on scientifically determined measured data.[334]

The comparison shows that the Hothouse diagram, like the diagrams of foxes and rabbits, represents a very simplified, more or less fantasized reality. The researchers of the Hothouse hypothesis realize this very well, as I noticed when I asked them for an explanation. One of the authors of their publications replied to me, "That's right. The figure is very schematic, and we have pulled it apart a bit to make the principle clear."[335]

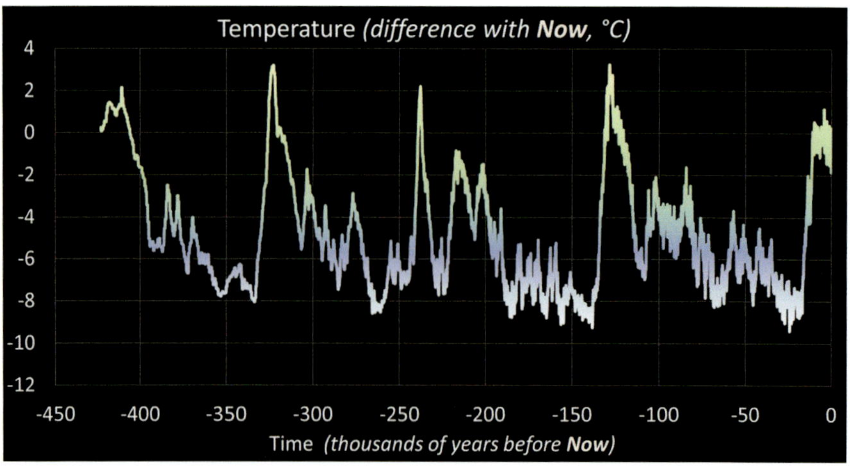

FIGURE 17.8 Temperature change during the last four ice ages.

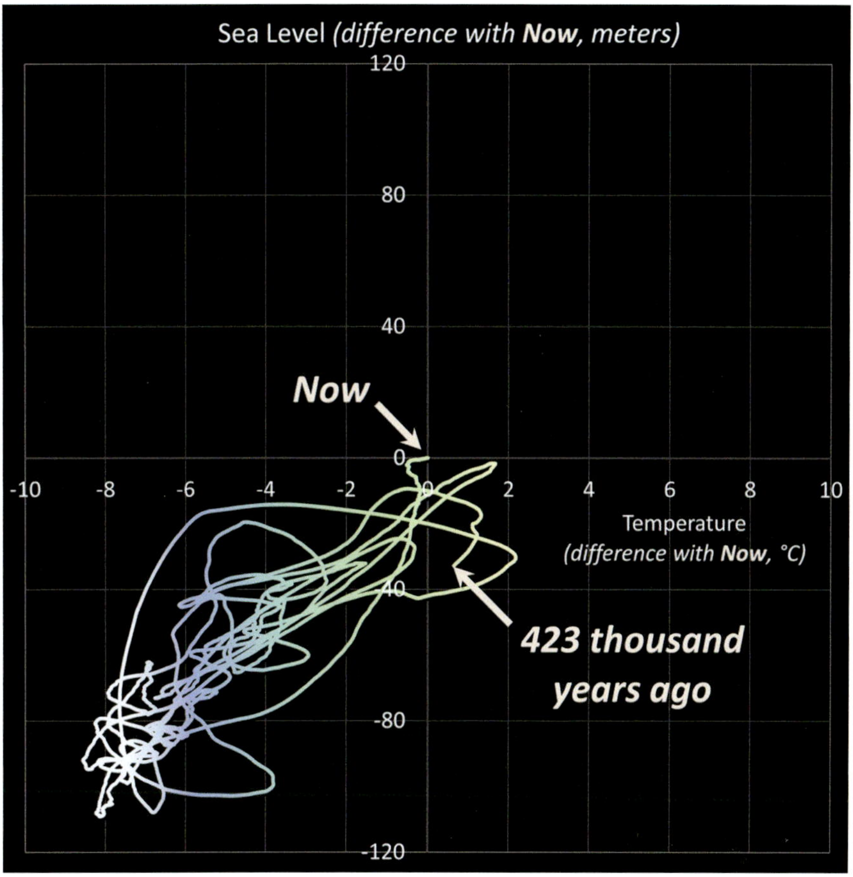

FIGURE 17.9 Ice Ages: phase diagram based on observed data.

The key question, of course, is the same one I posed in Chapter 7 (page 81) with respect to economic DSGE models: what good are such models anyway? The answer is: that depends on what you want to do with them. What you certainly cannot do with them is predict the future, that much is clear. Undoubtedly the future development of the climate will be at least as capricious as the past, shown in Figure 17.9. However, exactly as the reaction of a Hothouse researcher indicated, you can illustrate certain principles with it, and then draw lessons from it. For climate change, this provides a formidable warning: it is quite possible that by the time the global temperature has risen by 1.5 or 2°C, next it will skyrocket, with sea levels rising by as much as a hundred meters. It is impossible to quantify how likely that is.

Attractors in complex systems really exist, as has been amply demonstrated. The realistic phase diagram of Figure 17.9 illustrates that such attractors are sometimes poorly visible, due to all kinds of additional influences and noise. Nevertheless, it seems that the phase diagram testifies to a truly existing cyclic attractor that has controlled the temperature and other characteristics of our planet for millions of years.

Meanwhile, serious attempts are being made to approach the world economy as a complex system as well. Publications are written,[336] conferences are being held[337] and relationships are being established between complexity, economics and ecology.[338] Several existing economic theories have been cast in complex models.[339] Unfortunately, many of these studies make little or no use of empirical data, so the results tend to relate to reality in the same way the abstract phase diagram of the ice ages in Figure 17.7 relates to reality shown in Figure 17.9.

The mission of omniconomics has been formulated: design and build a global complex system, including an economy, that relies on equilibrium attractors.

The obvious question now is: how do you do that?

More new tools are available to help find the answer. Since human, technological and natural society is made up of billions of actors – from bacteria to chairs, people, companies and forests – interacting in a highly complex network, I turn to *network theory*.

Recommended further reading (for details, consult the online Bibliography)

Book: Lewin, Roger (1992): *Complexity: Life at the Edge of Chaos*.

Book: Strogatz, Steven (2003): *Sync. The Emerging Science of Spontaneous Order*.

Proceedings article: Steffen, Will et al. (2018): *Trajectories of the Earth System in the Anthropocene*.

Book: Hausmann, Ricardo; Hidalgo, Cesar A. et al. (2011): *The Atlas of Economic Complexity: Mapping Paths to Prosperity*.

18
TEGULAR UNIFICATION

The study of complex systems, the subject of the previous chapter, looks at reality in a new way, compared to traditional fields of science. These fields are used to examining a particular, identifiable part of the world. Chemistry studies the reactions between chemical substances; biology has living nature as its subject. Organization science wants to know everything about organizations, psychology about human emotions and behavior.

Complexity experts do things differently: they look at *all these areas*. They put on a different kind of glasses and emphasize not the features that characterize those separate systems, but rather the *similarities* between them. Complexity builds a bridge between apparently completely different fields of science by looking at them from a higher level of abstraction.

The same applies to another new perspective: network theory.

18.1 New equipment: network theory

Figure 18.1 shows two examples of networks. They look similar, although they have completely different scales: the global Internet is a billion times the size of a rat's brain. One was created by humans, the other by evolution. And yet the structures are similar. You can also find such structures on a metro map of Paris or in the finely branched vascular system of a human being.

There are other such networks, but they are less visible. Think of the network of your family members, friends and colleagues. The thought network I created as a baby, shown in Figure 13.1 (page 169); and the network of causes and effects in Figure 17.1 (page 230) that drive climate change. Like complexity, network theory has become important in almost every field of science.

DOI: 10.4324/9781003306269-20

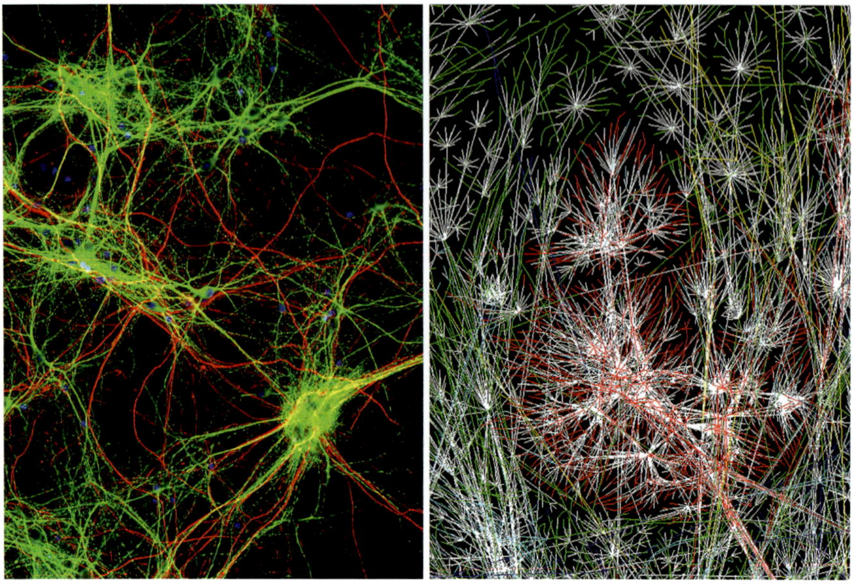

FIGURE 18.1 Two networks. Left: brain cells of a rat. Right: a part of the internet.

When a field of science involves only actors that are more or less the same size – all humans, or all brain cells – then research is still relatively easy. But when you deal with networks in which the elements are themselves networks, things get a lot more complicated.[340] The challenge of such multilevel networks has been taken up by quite a few research teams. You can think of research on organizations,[341] game theory,[342] urban areas, medicine, economics, finance,[343] ecology[344] and more.

Omniconomics makes the challenge even greater. When it comes to multilevel there, the aggregation levels are very far apart, at least from nanotechnology at level −3 to planetary conquest at level 4. Layer upon layer upon layer: atoms forming complex molecules, forming living cells, forming people, forming communities, forming global society, forming part of the entire Earth's biosphere, the most complex system we know.

Not at every moment will all those scales be relevant; but if the aim is to make one overarching *Universal Omniconomic Theory*, of course they are. And that raises two questions. First, can it be done, unite so many scales in one theoretical model? And second, should it be done?

18.2 Level transitions

When you board an international train to retrace the route taken by the famous explorer Marco Polo along the Silk Road to the Far East from the year 1271, you start from the wealthy European city of Venice. In the course of the journey, you'll pass through the crusader city of Akko in Asia Minor, the Turkish city of Trebizond and the Arab metropolis of Baghdad, among others. Via a boat trip to the Persian island of Hormuz you reach the Afghan city of Balkh. After that it gets tricky, because now you have to cross the Pamir high mountains by caravan to arrive in Kashgar, a city in the far west of China. This is followed by another difficult travel section through the Taklamakan desert, which takes you to the Buddhist city of Dunhuang. After a trip to the Mongolian capital Karakorum, you finally arrive in China's capital Beijing, as Polo did in 1275, a few years after that city was bloodily conquered by the Mongols led by Kublai Khan, who thus became China's first Mongol emperor.

During the long journey, a colorful collection of the most exotic cultures passes before your eyes. Colors, architectural styles, languages, myths: they differ enormously from place to place. Starting from a Christian part of the world you see successively Jews, Muslims, Tengriists, Zoroastrians, Manicheans, Buddhists, Lamaists and Shamanists, and in China: Confucianists, Taoists, Ancestorists and probably more, even if they are nowadays covered by a Chinese Communist veil. Just imagine the scents you'll sniff out! With the spicy Venetian memory of oregano and basil still in your nose, you experience the spicy vapor of coffee with nutmeg or cardamom in Akko and the sweet bouquet of fruit and caramel in Trebizond on the Black Sea. A thousand kilometers further on, you discover the scent of the deep red saffron of Persia, which now calls itself Iran, and the aroma of the hot-blooded ochre-yellow Afghan curries. At the end of the trip, you enjoy a Chinese dinner spiced with cinnamon and aniseed, concluded with a fragrant cup of jasmine tea. At least that's how my imagination tells me, and anyone who can improve on it may say so.

After this fascinating journey along a spatial dimension from west to east, I will take you on another train ride. This time we travel along the scale dimension from small to large. Let's board at level −5, at a station called 'proton', one of the most famous elementary particles. Before you leave, you look around you, seeing bewildering things. Shapes you cannot discover, nor sizes. Even places and speeds are hard to discern: the better you try to see where an object is, the more unclear it is whether it is moving and if so, how fast.[345] Rather, the objects appear as a kind of phantom, which the physicists call 'fields'.[346] It's like living in a mist.

Just as you will never find a Christian crucifix in a shamanic temple, you will never encounter the characteristic properties of the actors at the 'proton' level in everyday life. They have been given names such as *baryon number*, *spin*, *isospin*, *strangeness*. Smells, like nutmeg or mint? They don't have those here. One half of the actors consist of matter, the other of antimatter; when two opponents meet, a flash of light follows and they both cease to exist.

From this alienating experience you pass over a railroad bridge into a gap at level −4: an uninhabited space. You travel further and cover a considerable distance: a size factor of up to a hundred thousand to station 'atom'. There you make an important discovery, because you arrive in a totally different culture. The inhabitants of this level possess chemical properties! They are busy playing games with each other under the watchful eye of chemists: they stick together and form all kinds of molecules, as if they were families, after which they occasionally let go of each other to form new relationships. It's a rich culture here, because soon you discover that there are more than a hundred different kinds of atoms that make up millions of different molecules.

Now when you take the train again to even higher levels, you can continue to travel in different directions. You can go in the enchantingly complex direction of *life*, but you can also choose a more chaotic direction, for example, the one in which you go to see the structureless *gases*. It was that route that I once followed, where I made an amazing discovery.

Statistical mechanics

I remember it well, from when I was a first-year student in university, seventeen years old: the *Statistical Mechanics* lectures.[347] That subject was the first time I was introduced to a genuine *level transition*. If I'm not mistaken, it is also historically the oldest discipline to accomplish that: it dates back to the 19th century.

At that time, people realized that all matter is made up of atoms. Gases do this too: they are made up of free-floating atoms, or else of free molecules each made up of two or more atoms, such as the greenhouse gases CO_2 and NH_4. Those atoms or molecules possess properties such as mass, location, and velocity, which you can use to perform calculations using physical *laws of collision*, discovered by Newton, and the *Law of Conservation of Energy*.

Now atoms and molecules have no *temperature*. I do not mean that their temperature is zero (0 kelvin or 0°C), but that 'temperature' is simply not a property of them. The same is true of *volume* and *pressure*: these too are not properties of such particles.

But if you now put a lot of these individual particles – think of about 10^{24} of them – together in a vessel, then together they form a gas, which immediately assumes the volume of the vessel and thereby has measurable characteristics: pressure and temperature. Experiments over the past few centuries gave us the *gas laws*: one of them dictates that if you double the temperature of a gas, you thereby also double its pressure. But the individual particles *don't have* a temperature or a pressure at all. How can that be?

In searching for an answer in the 19th century, the recently invented statistics came to the rescue. When you combine the collision laws of separate particles with statistical calculations, you get outcomes, including certain *averages*, that turn out to behave mathematically exactly like the temperature and pressure of the gas. The calculations show in detail how the properties of the gas emerge from those of the individual particles.

As a first-year student, I thought that was an amazing discovery! This is an approach that shows how the *larger* can be fully explained from the constituent *smaller*. It is a brilliant example of a well-understood level transition and a great *reductionist* success.

But: this approach has an important limitation. Statistical Mechanics was so successful, because it deals with a situation in which all the (very many) lower-level actors are thrown together in a completely chaotic way into a higher-level vessel and nothing special happens. In that simple situation, statistics helps us to understand and even predict the emergent behavior at the higher level.

But now see how such a large number of particles can be placed together in a much more exciting way. To do so, this time choose, from train station 'atom', the more interesting upward track towards complexity and life. Notice how from atoms first small and then giant molecules are built: proteins, carbohydrates, fats. Enzymes, hormones. Even DNA strands, made up of thousands of atoms. Travel further along a series of higher levels and discover that with those molecules living cells are assembled, which in turn are the building blocks of even more complex bodies: an ant, a palm tree, a tiger. Or a human. We have now arrived at level 0 but we travel on lustily, seeing how individual plants, animals and humans are in turn the building blocks for jungles, cities and countries. And finally of the sizzlingly alive outer shell of the entire Planet Earth.

Hindsight reductionism

How successful is reductionism during this train ride into complexity? *Half.* The easy half, to be honest. When you look at a living cell, biological science is getting better at understanding how in that cell the molecules or even the individual atoms interact with each other. In the 21st century, we are working hard on that level transition, which is comparable to Statistical Mechanics, but focused on the infinitely more difficult issue of moving from inanimate atoms to living matter. The field of research is called 'quantum biology'. One of the researchers wrote, *"The biggest question [...] is how the inert atoms and molecules found in rocks are transformed every day into running, jumping, flying, navigating, swimming, growing, loving, hating, lusting, fearing, thinking, laughing, crying, living stuff."*[348]

That research is increasingly successful, thanks to the fact that you can see the end product of the building process: the complete cell. This is '*hindsight reductionism*': with the end result in your hands, you can look at the constituent parts and understand all sorts of things. But what does not work is 'foresight reductionism', that is: predicting from the properties of the separate atoms the characteristics of the living cell to be built with them, as long as that cell is not yet there.[349] It is just like with Lego constructions: when I see a large plastic castle, I can easily recognize the individual Lego bricks in it and discover how the building children went about their work. But if I arrive a bit earlier and only see a pile of loose bricks, I can't possibly predict the castle, because you can put the bricks together in an infinite number of ways.

The lesson is: when large numbers of small actors are assembled in a *chaotic* way into a larger whole (as in a gas), then you may apply statistics. You may not then be able to calculate the behavior of the individual actors (because of the butterfly effect, see page 230), but there is a good chance that you can predict the emergent properties and even the evolution of the larger whole quite nicely.

In contrast, when the actors are brought together in a *complex* way, totally unpredictable new properties emerge. Statistics are then inapplicable. This is all the more true, when multiple layers of complexity are stacked on top of each other: complex molecules with which living cells are assembled, which together form complex bodies of human beings that in turn together are a complex human community containing, among other things, a global economic system. Layer upon layer, complexity stacked on top of complexity on top of complexity.

18.3 The rooftile structure

Now, physicists have been searching for years for their 'holy grail': one *Theory of Everything*, also referred to as a '*Grand Unified Theory*'. The unification project is not going badly, as they have already succeeded a number of times in merging major theoretical constructs. For Statistical Mechanics, the level transition that connected atomic theory with gas and heat theory, is not the only success. Another 19th-century triumph was the unification of electricity, magnetism, light, radio waves and X-rays into one theory that together has since been called *electromagnetism*. The unifying work is not yet finished, as I already wrote on page 216: quantum mechanics and the General Theory of Relativity still refuse to become one.

In the meantime, physicists and chemists did manage together to firmly connect their fields. Biochemists are building bridges between chemistry and cell biology. And neuropsychologists are currently working on tight connections between biochemistry and psychology. Unification succeeds on many fronts.

Yet, despite that unification, all of these individual disciplines still exist. *Chemistry* has not become a part of *physics*. *Biology* has not become *chemistry*, *psychology* is not *biology*, *sociology* is not *psychology*. Partly this is for practical reasons: if you wanted to describe everything that happens, including sociological developments, at the atomic level, it all became completely unreadable. But a much more profound reason is that you would thereby completely overlook all the emergent properties of the various higher levels. All those fields, even after bridges have been built between them, still have their own words and narratives. Even their explanatory principles differ.

Economics: a level fallacy

Just as *physicists* studying individual atoms are not conducting *biological* research, *psychologists* studying individual people don't conduct *economic* research. The top level of the complex macroeconomic system, consisting of billions of people and trillions of other actors and objects (from screws to corn plants to websites to united nations), is much higher than the level of individual people. The giant economic system possesses – it cannot be otherwise – emergent properties that individual people, and even individual companies, institutions, or countries, don't possess at all. I suspect that we don't even have the slightest notion about many of these emergent properties yet.

Quite a few economists believe that economics is a human science. Others define it more broadly and call economics a social science. Both claims are a *level fallacy*. Just as biology is not physics, macroeconomics is not psychology or sociology.

This is the fundamental cause of the failure of economics as a science: the aggregation level is wrong.

Love

Reductionism and causality are two explanatory principles that have proven very successful in natural science. For example, the capacity to love can be excellently described reductionistically and causally, on the basis of *neurotransmitters* in the brain: on page 197, I mentioned anandamide and a few other substances. But those who see biochemistry as the only explanation, miss essential aspects of love, which are sung about by poets and artists and described by psychologists, who thereby adopt opposite explanatory principles that are holistic and finalistic, that is: with the whole human being as an indivisible person who strives purposefully for love fulfillment.

So, who is right: the causal-reductionists, or the finalist-holists? Who wins? The answer, of course, is that they are *both* right. Because they view human beings from different levels of aggregation, they are complementary. They need each other. Anyone who sees only one of the two explanations as the only correct reality is half-blind.

More generally, anyone who demands that every piece of reality must be fully understood in exactly one way, experiences paradoxes and impossible dilemmas in all sorts of places. Such as the paradox of free will, which on the one hand we must have, among other things because of emotional, legal and social considerations, and on the other hand we cannot have because of biochemical and neurological arguments (see page 48). Or the paradox of the 'I': is a human being one indivisible person or rather a kind of assembly or swarm within a single head (see page 170)? Earlier in this book, therefore, I argued that the current, paradox-provoking application of 'free will' and 'I' should be labeled as impetus words and redefined, or else replaced with other words.

The covering roof

Traveling from Venice to China, across Europe and Asia, you experience a rich collection of spicy scents, from oregano to cardamom and from saffron to jasmine. At the same time, you will encounter numerous religions during that journey. Each has its own explanatory principles, such as: gods and demons, good and evil, yang and yin. And their own narratives, varying from original sin to karma, from salvation to enlightenment, from afterlife to rebirth, and from fun with seventy heavenly virgins to nirvana, non-existence.

Anyone who puts all the European and Asian spices into one stew; or who tries to build one universal religion for all, out of all the world's religions, will at best succeed in making a stinky mixture that pleases few. What biodiversity is to ecosystems, cultural and religious diversity is to human communities: fragrant and indispensable.

Traveling from protons to human communities and higher, along the level dimension, one encounters numerous fields of science. Each has its own explanatory principles, such as: matter and antimatter, male and female, North and South. And their own stories, ranging from conservation of momentum to solidarity, from electrical to romantic attraction, and from supernovas to life fulfillment. Those who try to build one universal theory for everything out of all the fields of science will at best succeed in creating something superficial that is of no use to anyone. For omniconomics, explanatory diversity is fragrant and indispensable. See also Figure A22.1 in Appendix 22, which shows other common explanatory principles, including 'similarity', 'opposites', 'anthropomorphism', 'God', 'logic', and 'instinct'.

Unification, a conclusive coverage of the entire field of omniconomics, cannot therefore be obtained with one comprehensive theory, based on one set of explanatory principles. Instead, what is needed is a colorful collection of theories based on a variety of explanatory principles, words, narratives, points of view, and aggregation levels.[350]

Now, in addition to being *unifying*, also *complete* and *coherent* are necessary properties of omniconomics, I wrote in Chapter 16.

This *completeness* criterion can be achieved with a collection of aligned theories that together cover the entire omniconomic field of investigation, in much the same way as a collection of roof tiles that align with each other can form one watertight roof. The explanatory diversity required for this is achieved, not with unification by one *Theory of Really Everything*, but with *tegular* (tile-like) *unification*.

Coherence is achieved with the help of bridge-building-level transitions such as Statistical Mechanics and biochemistry, aided by cross-disciplinary tools such as mathematics, statistics, simulations, complexity and network theory.

Recommended further reading (for details, consult the online Bibliography)

Journal article: Menichetti, G., Dall'Asta, L. & Bianconi, G. (2016): *Control of Multilayer Networks.*

Book: Al-Khalili, Jim & McFadden, Johnjoe (2014): *Life on the Edge. The Coming of Age of Quantum Biology.*

Book: Durston, Sarah & Ton Baggerman (2017): *The Universe, Life and Everything... Dialogues on our Changing Understanding of Reality.*

19

ON THE ROAD TO SUCCESS

Evolution towards sustainability

And so, there are now a lot of new physical and mental tools available, which can be combined together to be good for step 1 and step 2 on the Path to Science. I have already mentioned complexity, network theory, simulations. There will be more in this chapter.

This has already led to new patterns (step 3) and new words (step 4), such as emergence and attractors, and to the new narrative (step 5) of tegular unification.

But these are all rather abstract results, which do not yet contribute to the replacement or improvement of economic impetus words and myths. It must not stop there, since omniconomics should not, as before economics, bubble up from the intuition of researchers but must take root in reality.

So now it is the turn of empiricism. Observations, that's what we need. Preferably: experiments, a long series of experiments. Repeated experiments even, both identical repetitions and with small or larger variations, because that's how science works. In the hope that the observations will give us new, realistic patterns, so that we can give those patterns new words and start telling new narratives with them. Followed by new policies and thus new effects.

But: there is only one reality. So how many experiments, how many iterations and how many variations on it can you do altogether?

There is a fantastic answer to that question, thanks to yet another new tool. Because today we have much more than one reality at our disposal.

DOI: 10.4324/9781003306269-21

19.1 New equipment: self-learning systems

I'm on the train. In First Class, of course, because I am an *Important Person*. Across from me sits Marco Polo. We are on our way to Beijing, but not right now, because the train is slowing down and letting out the steam whistle. I know why: we are approaching Karakorum main station.

I look out the window, where the landscape of the Mongolian steppes is slowly turning into inhabited territory. One *ger* after another, the famous Mongolian tent, comes into view. Hundreds of people, dogs, goats and camels walk among them. Fires are burning, cooking pots hang above them. Marco says something to me, but I don't understand him because I don't speak Italian. So I make a subtle twist in the air with my right hand and hear him say, "…from China." "Sorry, what are you saying?" I inquire politely. "That Kublai Khan reportedly emerged victorious yesterday in the election of the Emperor of China."

That is a remarkable event, because Kublai is not Chinese at all, I know. For a Mongol to win the Chinese election is an unprecedented event. What a beautiful, open country it is! I am very curious to see what the new Emperor will mean for economic structure development. For sustainability.

The doors of the train swing open and a strong smell of charcoal fires and simmering mutton reaches my nose. Those fumes are obviously coming from my snuff box, which is located in the lower part of my 3D gaming glasses. For I am not only in Karakorum but also in my own office, typing up this report for you. I am an *Important Person*, because I am a real, human player in an online game in which most of the participants are created by the game. Marco Polo is one such virtual human being, at least I suspect so. As is Emperor Kublai Khan, although you can never be sure, because the virtual persons in this game have passed the *Turing Test*[351] by a wide margin, that is, they are indistinguishable from real people. Real people are in this game in large numbers, because this is a 'MMORPG', a *Massively Multiplayer Online Roleplaying Game*. According to the game producer, there are 223,416 human players present in the virtual reality of this game at this time, along with about ten times that many virtual players. That is indeed 'massive'.

Although most of these approximately 2.5 million real and imaginary persons have no knowledge of it, we are participating not only in a game but simultaneously in an important scientific experiment, which is part of a chain of experiments that together form an omniconomic research that is going to change the world. For we find ourselves together in one of the countless virtual realities that have been brought into existence on the Internet.

The virtual reality in which I find myself is a MMORPG. OK, 'reality' is an exaggeration, because the MMORPG in which I am on the train and travelling with Marco Polo does not exist; I have fantasized about it for this book. But other MMORPGs do exist, and in their simulated realities a hundred thousand or more real people can be walking around at any given time – well, not they themselves but their avatars, their 'pawns' – where it doesn't matter where these people are in the real world. If you ever play such online games, you probably know *World of Warcraft*, and maybe also *Rift*, *Elder Scrolls Online*, *Guild Wars* and *Neverwinter*. If you're a little older, you might remember the virtual world of *Second Life*.

I'll say right off the bat: today's MMORPGs are not yet suitable for use as omni-conomic experimentation spaces. But something can be done about that. When scientists and game developers work together, games can be developed that do lend themselves to this. Their virtual worlds must then meet a number of requirements. They must be sufficiently *complex* to allow room for surprising *emergent* developments and the emergence of *attractors*. They must be sufficiently *layered*, in the sense of an accumulation of *aggregation levels*. This takes an enormous amount of computing power, but that is not a problem when you have a network of hundreds of thousands of computers. Moreover, both requirements, complexity and layering, do not have to be realized all at once: if current trends continue, they will come about in the next few years.

Artificial Intelligence

But that's not all. Another important prerequisite is that the games, or the virtual worlds in general, become increasingly intelligent. Then, of course, I'm talking about *Artificial* Intelligence (AI). In the oldest forms of AI, all 'cleverness' was programmed in by humans. But those possibilities were very limited: it was as if, as young parents, you wanted to give your children's entire upbringing and schooling program to them ready-made during conception, so that they would be born as 'small but completed adults'. Just as children (and adults) learn constantly from their experiences, so do modern AI programs. If you want to teach a computer to recognize dogs, you do so not by using a formal definition of a dog (which is really not easy, try it…) but by offering numerous examples of dogs and of things that are not dogs. "That's a doggie. No, that's not a doggie, that's a cow!" (I can still hear myself saying it to my toddler son. "What does a cow say again?")

Huge advances are being made with such self-learning systems. The most powerful systems run not in traditional computers but in neural networks of interconnected computers that together resemble human brains in structure, and are designed with the help of network theory. A sensational example of a self-learning system is the development of the 'Go' computer.

Go

During my university studies in astronomy and physics in the years beginning in 1972, I was an avid but moderate club player of the mind sport Go, which is played with black and white glass stones on a 19-by-19 board. The sport is much more complex than chess, and it is no wonder that after the first victory of a chess program over the human world chess champion, it took almost twenty years for the same to happen with Go. 1997: *Deep Blue* against world chess champion Gary Kasparov, a match over six games; result 3½–2½. 2016: *AlphaGo* against world Go champion Lee Sedol, a match over five games; result 4–1.

Deep Blue was not a self-learning chess program, it was based on *GOFAI* (Good Old-Fashioned Artificial Intelligence); that is, on brute force: it knew an incredible number of games by 'heart' and could calculate a number of moves ahead amazingly quickly. Critics did not view this as intelligence at all.

AlphaGo, on the other hand, learned its mind games by playing them itself, losing a lot in the beginning but getting stronger as it went along: a convincing example of real intelligence. The news of *AlphaGo's* victory made headlines all over the world.

After *AlphaGo*'s victory, something strange happened: the program was immediately retired, to make way for *AlphaGo Zero*. Whereas the original *AlphaGo* was trained by playing against humans, *Zero* was only given the rules of the game and nothing else: it had to get by, by playing games *against itself*, with the help of a computer power that was only one-tenth of that of his predecessor.

After three days of practice, *AlphaGo Zero* had mastered complex game techniques such as might be used by grandmasters. After forty days, the program was applying successful strategies that had never been devised by humans in the thousands of years the game has been in existence.

In October 2017, a year and a half after the 'old' *AlphaGo* won four out of five games against the human world champion, *AlphaGo Zero* challenged its digital predecessor in a match over a hundred games. *AlphaGo Zero* defeated *AlphaGo* by 100–0.[352]

Reader, I don't know how you experience this, but this news item still gives me chills.

It didn't stop there. A subsequent version of the self-learning program, christened *AlphaZero* for short, taught itself from scratch a year later in thirteen days of practice to beat *AlphaGo Zero*, after which it took that same program nine hours to teach itself chess and overcome the reigning computer chess champion and then twelve hours to do the same with shogi, also known as 'Japanese chess'.[353]

Moravec's paradox

AlphaZero is an overwhelming success of AI. At the same time, however, other successes are taking longer than many expected. The self-driving car, for example, which after pre-programming must learn to drive in heavy traffic on its own, is still a disappointment. In old predictions of the future, the translation computer was also expected much earlier, with a program that allows you to talk in real time in other languages or at least understand them, just as I was able to converse directly with the Italian explorer Marco Polo after turning a virtual knob in the MMORPG I fantasized about. We are far from that point, however, although online translation programs such as *Google Translate* and *DeepL* are making rapid progress.

It seems rather contradictory: what we as humans find difficult, a computer program can learn smoothly; but what we do easily and automatically, often even unconsciously, is actually very difficult for computers. This is Moravec's 'Paradox', named after Hans Moravec, who articulated this contradiction in 1988.[354] One reason is: complexity. The game of Go may be the most difficult mind sport we know, but the world of the Go board is essentially very simple. But traffic is a particularly complex environment. And to translate texts correctly, you need to understand what you're translating, to avoid weird blunders, and for that you need to be familiar with the real, complex world of humans. The same is true if you want to artificially mimic people in a MMORPG in such a way that they pass the test, proposed in 1950 by computer pioneer Alan Turing, namely that you can no longer determine who are the mimicked people and who are the real ones. My fantasy about the train to Beijing is therefore rather ahead of its time, for in it, Marco Polo and Kublai Khan passed the test; but we are yet far from that point.

Wonderful results are already being achieved with complex self-learning systems. On page 209, I mentioned the example of the painkiller PZM21, designed by a computer program. In 2019, it was announced that another AI, based on long lines of astronomical observations but otherwise unaided by humans, independently discovered that the planets revolve around the Sun in elliptical orbits.[355] In the medical field, neural networks are becoming increasingly important in interpreting scans and determining diagnoses.

Meanwhile, AI systems are operating with growing success in game worlds that require far more complex and human insights than Go. In 2019, the *AlphaStar* program managed to become champion in the *real time strategy* (RTS) game *StarCraft II*, considered one of the most difficult adventure computer games.[356] In the same year, *OpenAI Five* defeated the human champions in devastating fashion in the game *Dota 2*, a strategy game in which you must work as a team to destroy the opponent's tower with a gang of fantasy characters.[357]

Virtual experiments

MMORPGs like *World of Warcraft* are, just like drug development programs and *AlphaStar*, computer simulations that mimic facets of the real world. I wrote already: simulations have become indispensable tools in all branches of science. Without computer simulations, most sciences would be nowhere near where they are today.

One of the earliest comprehensive simulations was Meadows' *World3* model (see page 98 and page 133), whose cautionary scenarios were an important link in the introduction of sustainable development. For omniconomics, simulations will be decisive, as I will show in this chapter.

MMORPGs, or more generally *virtual reality*, are a special class of simulations, because they visualize the simulated reality and make it possible to evoke the feeling of being within the virtual reality and actively participating in it. This has a huge advantage, because it makes it possible to conduct scientific experiments in which real people act in large numbers, if desired without them realizing at all that they are part of an experiment. In doing so, you can prevent the experiments from being based on a variant of *homo economicus*, producing results that have little to do with reality.

Virtual realities usually take the form of online games. The games industry is huge: global sales in 2020 reached $180 billion, far exceeding the entire film industry, which had sales of $101 billion in that year.[358] The market for virtual worlds has been growing at a rapid pace for years and will continue to do so for the foreseeable future, as there are many more possibilities than just games. Just think of holiday travel: wouldn't it be great to visit Beijing as a tourist without the tiring and environmentally damaging flight, either by just being there in zero seconds, or via a train journey like mine with Marco Polo? Or think of feature films or books in which you can be present yourself, as an extra or even as a main character. *"The tenth member of the nine companions of the Fellowship in Lord of the Rings!"*, my wife enthusiastically suggested. Other possibilities: traveling through the solar system, through history, through erotic or other fantasies, or through pristine nature. Trading, with or without money. Meetings. Switching genders for a while. Flying. Being healthy (if you're not in the real world). Being famous. Meeting God. Being an animal. Meditating: being a tree. Or a rock.

Simulations, whether in the form of virtual realities populated by real people or not, make it possible to conduct unlimited omniconomic experiments. The advantage of experiments without real humans is that they can be performed very quickly, not being slowed down by these biological creatures. (In human–computer interaction, for example in writing this book, the latter typically spend more than 99% of its time figuratively drumming its fingers impatiently – with the exception of the graphics of modern games, which take up a lot of the video chip's time.) Therefore, I suspect that the most effective strategy is a combination of both kinds of simulations: huge numbers of fully digital experiments that explore all kinds of possible worlds at lightning speed, plus a smaller number of virtual worlds in which real people are active, which allows researchers to search for features of intrinsic sustainability based on equilibrium attractors.

19.2 New narrative: evolution towards intrinsic sustainability

A little critter is crawling across my screen. It even crawled into my book, sorry about that: now it's in your copy too. It consists of a sort of head with a long, fire-red tongue and a pair of eyes, and further a body and some legs of varying length. It crawls around and leaves a kind of blue-grey slime trail. I have to admit: it's my fault, because I made the creature. And yet I didn't. Here's the thing.

I thought it would be fun to write a computer program in which evolution occurs. Real evolution, through a genetic encoding of the characteristics of a creature that reproduces. The reproduction process is not perfect, just like in real life, because errors occur in copying the genetic codes, *mutations*. (*Everyone* is a mutant, you too.) If I then make sure that variations in the little animals occur based on those errors, then there is a possibility for natural selection based on *survival of the fittest*.

Anyway, so I started programming. In order not to make things too difficult for myself – computer programming is just a hobby of mine, I am not a professional software developer – I chose to fixate the basic building blocks of my animals and thus not let them evolve: head, trunks, legs and eyes.

Evolution would relate solely to the way these Lego bricks are linked together and the way they hinge on each other. I called them *genimals*, short for genetic animals.

I began by designing the genetic code, which looked like, for example:

> 9D539A59C9218EA271EEBA47F62EF7F6CA17E27723FC782693 5EF8C8.

I will not reproduce the details of the encoding here: you can find them in Appendix 28. Using this code, I fully defined both the physique and the behavior of the genimal.

Next, a 'natural environment' was needed, a habitat, in which the genimal could spend its existence. In that habitat, the animal then had to perform a certain task in order to survive. Therefore, I created a green Garden in which I planted ten trees. The task of the genimal was now, to reach those trees one by one and to eat them bare: see Figure 19.1. For that purpose, I gave the genimal a limited life span. When that was over, the dying animal reproduced itself and passed away.

Reproduction, in my program, is efficient but unexciting: just a matter of copying and pasting the genetic code. With a deliberate but random error, making the child just slightly different from the parent. I released the child into the Garden, after which it was allowed to try to find trees just like its parent. If the child succeeded better than its parent, then the child won and was allowed to reproduce in turn while it died; if not, then the parent got another chance to reproduce. This meant survival of the fittest in the form of competition between parent and child. In time, I reasoned, more and more effective and smarter critters would emerge. Voilà: evolution.

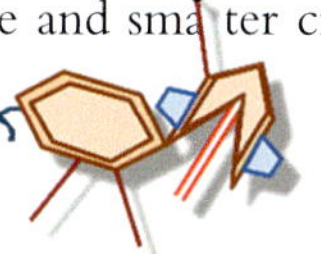

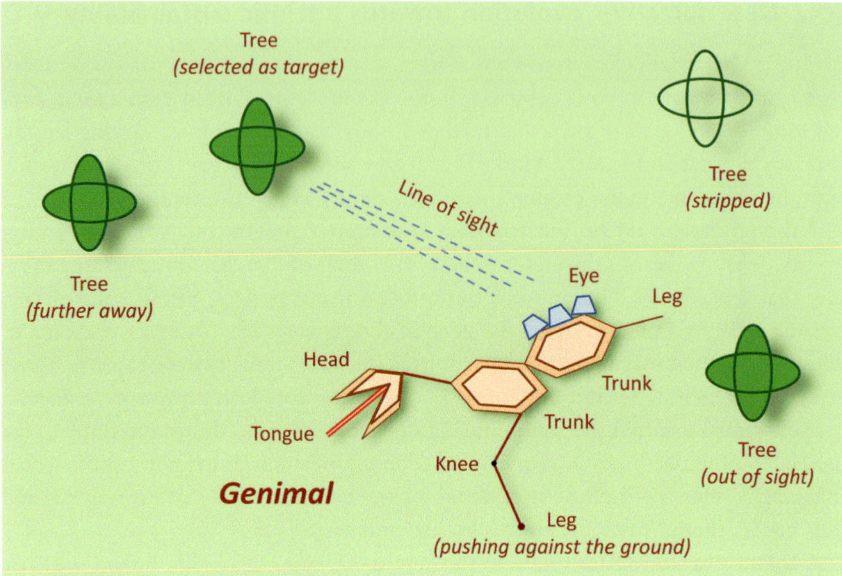

FIGURE 19.1 A genimal and some trees.

Whether it worked? You bet it did! I instructed the program to start each new evolution with a genimal constructed with completely random code. Such a null genimal is therefore, as you might expect, utterly clumsy in terms of physique and behavior and gets nothing done. But after a varying number of generations, a certain dexterity and efficiency emerges in some evolutionary processes, which after hundreds, thousands or tens of thousands of generations leads to the smooth, in the long run even increasingly smooth, detection of all trees. As in real nature, different animals discover totally different strategies for moving around: some skip along, others row or wobble, yet others whip around wildly.

By running the program many times, a varied zoo emerged: see Figure 19.2, in which I used my imagination to give the animals names based on their body shape and movement patterns. A number of characteristic movement patterns can be seen in Figure 19.3.

You can watch for yourself how these genimals evolve, for I have posted some video clips of them on YouTube, with a link on the website of this book. In fact, you can follow the evolutionary process all the way through, as the computer program 'Genimal' is also online available for download.

Artificial life

Genimal is a typical example of *artificial life*, a category of computer programs often referred to by the abbreviation *A-life*. The computer models of foxes and rabbits used to create Figure 17.2 through Figure 17.6 are also examples of A-life. Genimal

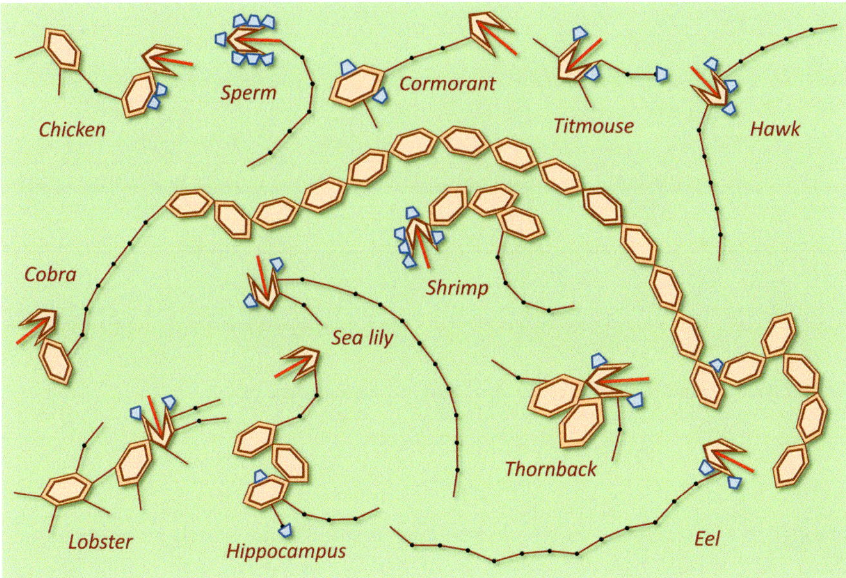

FIGURE 19.2 Genimal: the Zoo.

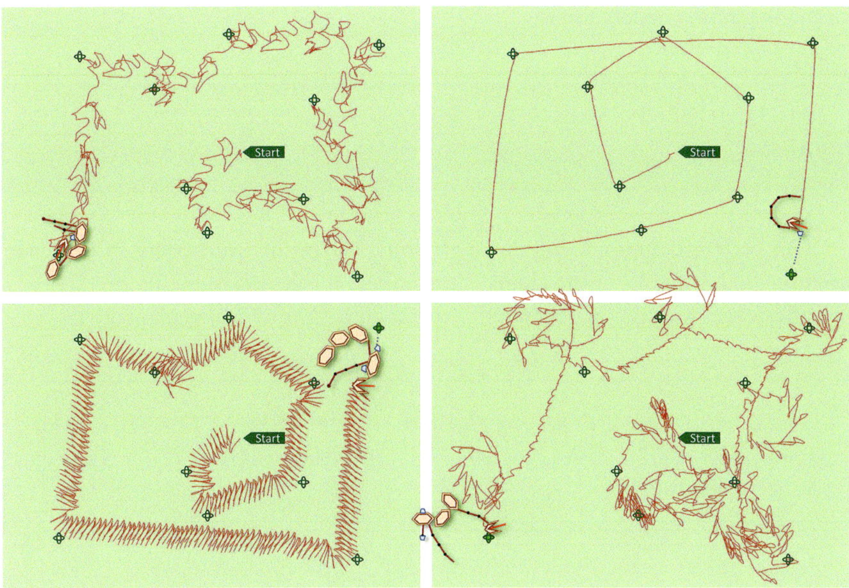

FIGURE 19.3 Genimal: movement patterns.

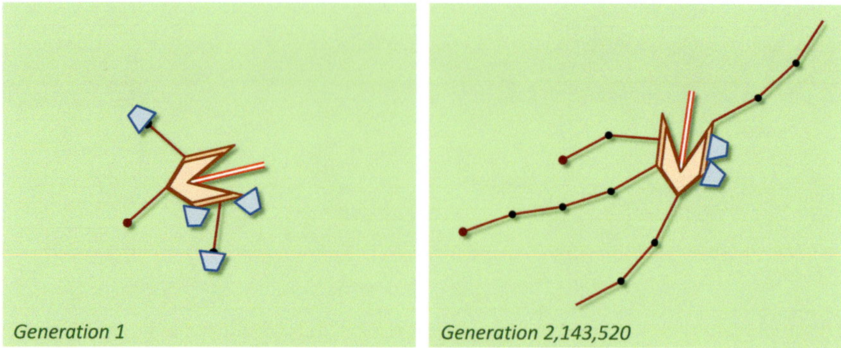

FIGURE 19.4 Evolution of a genimal. Right = a descendant of left.

belongs to a special class, in which evolution is simulated. In such programs something very special happens, which is of decisive importance for omniconomic experiments.

Genimal is a computer program. But the genetic coding of my genimals is a computer program within a computer program. Consider the example in Figure 19.4.

In that picture you see two phases of one evolutionary process. The genetic code of the beginning generation is:

```
F4CD29DE2839555464C2A407C27AEA6FF112EC8A729A32F3C6
AE266C.
```

This code is composed entirely at random. When the instructions in this code are executed, it results in the clumsy little animal on the left in the figure.

After over two million generations, the code has evolved incrementally to:

```
29965058C292A3EFD81967C047DE2AFB459F36F7C6D77F7C1586
C33C1586C33C27A423C3071BDDD52B12FDD3FE1DF5957C
0C58ED8D2BC561.
```

This code produces the graceful and skillful genimal visible in the right half. Both codes contain a series of prescriptions for the construction and behavior of the genimal, just as the DNA in your body cells prescribes how you grew from fetus to adult, and even how you are broken down again, for your natural aging is also recorded in your DNA, as is your vulnerability to disease and malfunction.

Both your DNA and the genimal encoding are sets of instructions that you can think of as computer programs. Only: *these programs were not written by anyone*. They arose spontaneously through evolution. Even though I created the program *Genimal*, I did not write the code of the successful genimal of Figure 19.4. In fact, I wouldn't even know how to do that.

Evolving A-life programs are thus self-learning systems. So are *AlphaGo Zero* and other artificially intelligent computer programs, but in a different way: they become increasingly powerful through their own experiences. In evolving A-life, individual generations learn nothing: development takes place when new generations do better than their parents, just as in real biological evolution.

Therefore, this is what I envision as the 'natural' environment for omniconomic experimentation:

> *Virtual worlds that are in their entirety giant living and evolving organisms, resembling real biological ecosystems, in which actors occur such as: real and virtual people, animal and plant species, enterprises, production processes, associations, educational institutions, friends and families, and so on and so forth.*

Such a habitat is a sort of living animal, whose characteristics change spontaneously over time through evolution.

One more ingredient needs to be added to this narrative. For: how do you ensure that such an evolving virtual world moves toward intrinsic sustainability, and not toward totally different, disastrous virtual worlds of blood and ruin?

19.3 Survival of the sustainablest

For the solution, I'll look again at the *Genimal* program, because in it I dealt with a similar problem. I told you that the job of the genimals is to detect and consume trees. But that's a little too simplistic. If that were the only success, a genimal which comes close to finding its first tree without actually getting there, would have accomplished nothing at all and thus would not have a better performance than its predecessor. That is why I have made it a little easier.

Success scores

Each genimal is given a score at every point in time. That score is composed of three elements: *A*. the number of trees found; *B*. the proximity to each tree not yet eaten; and *C*. the time used to find trees. The first number counts very heavily in the score; the second much less so. However, the used time provides deduction of points, so those which are faster are rewarded.

As a result of this success score, a child can beat its parent (or vice versa) without finding even one tree, simply by moving just a little bit more and thus slightly reducing the proximity to a tree. Thus, survival of the fittest can take place in the form of baby steps.

You can symbolically represent the success formula of the *Genimal* program as:

$$Score = A \times trees\ eaten + B \times proximity\ to\ trees - C \times time\ duration$$

In this formula, A, B and C are the weight factors that determine how heavily the various criteria count. A is much larger than B; note the minus sign in front of C. By choosing the criteria and the sizes of the weighting factors, you define what 'success' means in this context and thus steer the evolution.

The actual calculation of scores in Genimal is a bit more complicated; you can find all the details in Appendix 28.

Human Development Index

Success scores are used more often. For example, sustainability science uses the *Human Development Index* (HDI), a measure of the level of well-being in a country, whose score is calculated with:

$$Score = A \times Life\ expectancy + B \times Years\ of\ schooling + C \times Income$$

Details of the calculation of the HDI score and a few other examples of success scores are available in the second and third sections of Appendix 28.

Survival of the sustainablest

Back now to the virtual experimental worlds of omniconomics. You can perhaps imagine how success formulas are also designed for such worlds, in which the various facets of sustainability are added together with carefully determined weighting factors: the resilience of nature, prosperity, the degree of freedom and security, the rule of law, you name it. Negative values such as violence, war, extinct plants and animals, crime, poverty and economic disaster count with a negative weight.

How does such a formula come about? What criteria count, and with what weight? This is a very important question, because the answer will determine the kind of world that will emerge through evolution. That answer will have to come from a well-structured interplay between science and society. I devote the next chapter to that topic, so I'll leave it for now.

Just as biological and artificial evolution are the result of *survival of the fittest*, so sustainability valuation in omniconomic evolution produces a result I describe as:

survival of the sustainablest. The virtual experiment spaces constantly mutate spontaneously and randomly. Because always the version with the highest sustainability score is allowed to reproduce, the virtual world becomes more and more sustainable. *Intrinsically* sustainable, because: as soon as damage occurs, either via a furious tipping point or stealthily, the score drops dramatically and the failing world disappears from the evolutionary tree.

Observations and pattern recognition in virtual worlds

So: virtual worlds will become more and more sustainable. Fine, nice. But what use is that to us in the real world?

To find the answer, a funny new kind of science is needed, one that conducts empirical research in virtual realities. For the new, intrinsically sustainable virtual worlds are not created by us: they evolve spontaneously, like the genimal of Figure 19.4. Just as I cannot design the code of that creature, it may well be that the evolved virtual worlds are indeed very sustainable *but we don't really understand why.*

It's like the development of your child: you bring it to life; you raise it to independence in good conscience; but in the end it goes its own way which you may not always understand.

We undoubtedly know the outcome of the sustainability score of each of our virtual worlds. Moreover, we can look at such a world from the outside. We can even enter it, walk around and be active in it. But the virtual environment may be so complex and layered that we cannot easily see which mechanisms cause such a high degree of stable sustainability. What new, never-before-seen properties have emerged? Where do the equilibrium attractors come from? There is peace, there is little crime, nature is doing well, but why exactly? How do we capture all that in theoretical models? Every new world creates its own biology, sociology, and economics: a delightful new field for researchers.

If that happens, we will have to resort to an entirely new method: *virtual empiricism.* That is: forget that this is a virtual world. Look around you. Make systematic observations, exactly as in the real world. Look for regularities and patterns. Create hypotheses, models and theories to understand the new worlds we have brought to life. Build science.

Next: compare the scientific models of the various virtual worlds and try to draw general conclusions.

If you can do that, the next step is: figure out how to apply those theories and conclusions in real – not virtual – reality.

Let's experiment!

Summarizing: let software developers work with biologists, ecologists, psychologists, economists and other researchers to design such evolving worlds. Worlds that are initially in a primitive stage of sustainability, just like our real world. Let them evolve. Conduct widely varying series of experiments. Repeat them. Introduce variations and see what that conjures up.

Some of these worlds are populated with real people. In others, everything and everyone is virtual. Some worlds possess or design money or some other medium of exchange. Others may not.

With money

Worlds with money like ours. Made of shells, of salt, of sheep, of gold and silver, of tin, paper or plastic. Money that you can touch or money that is just a number on a screen. Money that loses a percentage of its value every year as determined by laws or administrative measures. Or that, on the contrary, gains value over time. Or money that can be passed on to another person a maximum of three times. Or one time.

Money that is subject to a maximum per person. Money with a constant total quantity: a real *Law of Conservation of Money*. Or money that has a fixed total flow, so that what one person receives more, is subtracted from what others receive: a *Law of Conservation of Money Current*. Money where the lowest paid get a legally determined percentage of what the highest paid get. Money that, when someone pays with it, does not make the payer's wallet emptier, just as the transfer of knowledge does not make you dumber: an alternative *Law of Conservation of Money*.

Worlds where the word 'money product' is considered a contradiction because money is not a product but a fiction, and where derivatives are therefore prohibited. Where making money through money trading is defined as a crime and punished accordingly.

Worlds where products cost money but services are free. Worlds in which the reverse is true. Worlds where all businesses are owned by the workers.

Worlds in which GDP is calculated arithmetically, or geometrically or harmonically, or perhaps all three: see page 183. Where GDP is calculated by adding the constituents not directly but weighted, where the weights are determined by social and environmental utility or harm, so that GDP becomes a true *success score*.

Worlds where everyone has a basic income. With a government that does not produce annual accounts, so that the treasury is an opaque pit, a black box, which means that the government has an endless supply of money at its disposal and therefore never has to borrow through government bonds or otherwise and can thus stimulate or slow down the economy without limits. In which the government never levies income tax, VAT or pension premiums. And pays no pensions, nor unemployment, sickness, invalidity or maternity benefits, because they are all superfluous thanks to the basic income. In which, of course, everyone is insured against medical expenses, fire, theft, vandalism and more. Even better: in which medical and other care is free, as are all other basic needs such as bread, the Internet, rice, sex, vegetables, literature, music, clothing, transportation, schooling, fruit and housing. Where poverty is impossible, so that the officially established human rights are real rights that you can claim.

Worlds where the less capital you have, the more votes you have in elections, or even more generally in determining policy, so that the rich are not simultaneously the powerful. A world in which money does not attract money but repels it, so that the rich no longer automatically become richer.

Without money

Worlds in which the production and distribution of food and useful goods never has to be halted for lack of money.

Worlds where those products are freely available, so no one takes more than they need (because you can always take it later), whether or not inhibited by social control, supported, or perhaps not, by digital tools and Big Data. In which, therefore, abuses are undoubtedly made, but perhaps less so than in a money economy so that crime decreases and power is more fairly distributed.

Worlds where there need not be, pardon me, where there *cannot* be, a distinction between economic in- and externalities, so that there is more natural respect for values that cannot be expressed in money since nothing is expressed in money anymore.

Worlds where stock markets no longer collapse, because there are no stock markets. Worlds where unpleasant work simply isn't done anymore because it isn't paid for, and where all indispensable work is thus made pleasant (or performed by computers and robots) because otherwise it simply doesn't happen. So that work is no longer experienced as work but as a hobby or passion.

Or worlds even crazier than I can think of. But you might.

And who knows

Worlds in which not only emperors and kings have become redundant and thus extinct professions, but also presidents, ministers and CEOs, as society has evolved into as mature a complex system as a bunch of brains, an anthill or a swarm of bees, which likewise have no leaders and in which no money exists since the individual brain cells and insects do not pay each other for services rendered but simply share everything. In which there is no anarchy but a natural order. In which democracy and societal participation are obvious for everyone.

Trial and error

Of course, it is likely that many of these fantasy worlds will fail hopelessly. Because those worlds will rapidly head for irreparable global bankruptcy, a destructive 'cultural revolution', a world war or even the rapid death by starvation of all those present: game over. Other worlds fail because they keep tumbling over the edge of chaos via tipping points, just like our real world, and never discover an evolutionary path to sustainability.

But that doesn't matter, because it's all virtual: it doesn't destroy anything in the real world. In the meantime, through trial and error, we learn at breakneck speed from both the failures and the successes. Sooner or later, intrinsically sustainable virtual worlds are going to emerge here and there, just as super-smart Go programs and nimble genimals have evolved, allowing us to discover more and more about what doesn't work and what does.

Is that a crazy ambition? Can it never be? I remind you of what I wrote on page 37.

Roorda's First Law: *Only when it succeeds do you know whether it is possible.*

Recommended further reading (for details, consult the online Bibliography)

Book: Moravec, Hans (1988): *Mind Children. The Future of Robot and Human Intelligence.*
Book: Mercado, Ruben (2021): *Artificial Economics.*

20

PARTICIPATORY DEMOCRACY

Society directs science

Success formulas. These must be designed so that they will determine the outcomes of *survival of the sustainablest* and thus guide the evolution towards intrinsically sustainable worlds. Presumably, there will not be just one formula, but a colorful variety that can be experimented with.

Who should make these success formulas? Scientists, of course. In close consultation with the software developers who design the virtual worlds. But that quick answer is not sufficient.

Just think about it: such a success formula consists of a combination of facets that together determine what we mean by sustainability. These include: a healthy climate and resilient nature; peace, health, education and prosperity for all; a stable economy, freedom, human rights and much more. If you perhaps even want to put all kinds of details into the formula – the conservation of each individual species, or the separate health care of each age group and gender – then it gets very complicated, since all those facets, with appropriate weight factors, have to be compared and related to each other. In itself, that is the work of experts, but: who chooses the themes? Who determines what weights those themes get? Actually, the underlying questions are:

What kind of world do we want to live in? And who gets to determine that?

Those who get to determine that, are they perhaps the scientists? That seems to me an inadequate answer. For, in essence, some of the issues at stake are centuries-old and are much more than just scientific questions. There are also philosophical, political, ecological, historical, emotional, humanistic and religious aspects. So, who decides: the philosophers perhaps, as the Greek philosopher Plato once suggested?[359] The priests, the imams, the rabbis and the lamas? The politicians, whether democratically elected and controlled or not? The trade union movement?

DOI: 10.4324/9781003306269-22

Concerned citizens? Action groups related to nature, environment or human rights? A world government – which doesn't yet exist?

20.1 Support

The answer is: yes, all of them. And everyone else who wants to, because the world belongs to *all of us*. The question is, of course: how do you do that, involve *everyone* in making choices about how to organize the world?

One obvious answer is: democracy. On this subject, Winston Churchill famously said in 1947, "*Indeed, it has been said that democracy is the worst form of Government except all those other forms that have been tried from time to time.*"[360] From the context of his statement, it is clear that he was referring to parliamentary democracy, based on majority decisions. Well, he was wrong. Because there is a better approach: decision-making based on broad support, wherever possible even: consensus.

Consensus

As a consultant, I have conducted many assessments related to sustainability in education[361] and business.[362] In such an assessment, about fifteen people are gathered for a half or a full day, and under the direction of the assessor, a wide range of topics are worked through. Together, the participants are representative of all kinds of stakeholders in the organization being assessed. This includes at least: management; employees (from high to low in the organization); and customers (or: clients, patients, students). If relevant, there are also external advisors, shareholders, inspectors or financiers.

During the assessment, opinions are exchanged by the participants. This is done on the basis of complete equality, with the executives usually being the last to be allowed by me to contribute their views, as they are the first to do so on all the other 364 days of the year.

No objective decisions can be made on many subjects, as they are, by definition, subjective. They are discussed one by one; together, the participants come to intersubjective conclusions: first about the current situation, then about a desired situation that is to be achieved in the course of one or two years. The most important rule is: *There is never a vote.*

So how do the decisions come about? On each issue, we keep talking and listening until everyone understands each other's points of view, all relevant data and feelings have been exchanged, and, finally, there is a broad base of support for decisions that are then taken by consensus.

Majority decisions

There is a good reason why, when I designed the assessment methods, I introduced this fixed rule: *never vote*. Because voting procedures have winners as well as losers.

The Vendelar family lives in Bloomington, Indiana: daddy, mommy and the children Hailey, Herbert and little Huey. They have a tradition: every year the children choose the destination for the family's summer vacations. The decision is made entirely democratically, that is, by voting. Hailey and Huey love the sun and the beach, so they always vote for Puerto Rico. Herbert can't stand the sun and would die to visit the Alyeska Resort in Alaska.

Where do you think the Vendelars will go again this summer?

To investigate whether such a scenario actually occurs, I looked at my own country and counted the Dutch cabinets. Since the Second World War, there have been thirty (reference date: 1 July 2021). I classified the political parties according to five main streams. (You can download the details as a spreadsheet called *Dutch Cabinets, 1945–2021*.) That produced the following result:

	Green	*Socialist*	*Christian*	*Liberal*	*Far right*
Prime ministers	0	8 *(27%)*	19 *(63%)*	3 *(10%)*	0
Coalitions	1 *(3%)*	14 *(47%)*	27 *(90%)*	21 *(70%)*	2 *(7%)*
Coalition days	1683 *(6%)*	15245 *(55%)*	23059 *(83%)*	20376 *(73%)*	1062 *(4%)*

The survey shows that the Christian parties provided the prime minister for 19 of the 30 cabinets; and that they participated in 27 of those 30 cabinets, governing 83% of the time. That's a lot: the numbers suggest that, like some of the Vendelars, the Christians may be getting their way quite often. That appears to be true, because: from the Second World War onward, they were a continuous part of the government for half a century. As a result, it was not until 1994 that the country was able to settle two human rights issues with which the Christian parties had strong ideological disagreements: same-sex marriage and euthanasia. The Christians, and to a lesser extent the Liberals, are the 'Hailey and Huey' of Dutch politics. The one-time participation in government of the Greens ended in 1977, almost half a century ago: they are 'Herbert'; they rarely get their way, which perhaps explains something of why the Netherlands is by no means at the forefront of policy areas such as climate change when compared with other EU countries.

Somewhere in Europe – it could be Asia – lies a very small country, with now three hundred inhabitants. In the year 998, a law was introduced there that stipulates that for every party that is part of a government coalition, in the next election the number of votes cast for that party has to be multiplied by 0.8, in order to moderate its power. If, in spite of this, such a party immediately returns to government, then the number of votes is multiplied by 0.8 × 0.8 at the next elections, and so on, until the other parties also get a chance.

The referendum

That law in that tiny country, of course, is just a stopgap. If you really intend to involve everyone in the process of thinking about and deciding in what kind of world we want to live together, majority decisions are useless. This is especially true of the ultimate majority decision: the referendum.

> **The European Constitution (which has not yet come to fruition)**
>
> On 18 June 2004, the heads of government of the EU agreed on the introduction of a European Constitution. It was a unique step in the history of Europe, which had been plagued by literally thousands of wars since the European Migration Period ended, around the eighth century AD.
>
> The Dutch government embraced the plan. In October, the Prime Minister signed the European Constitution at a ceremony in Rome. This was done with the support of a large majority of the Dutch parliament: 127 members of the House of Representatives voted in favor, and only 22 members voted 'no'.
>
> In the meantime, a law had been passed in the Netherlands, stating that ratification would not take place until the people had voted on the Constitution in a referendum. This was an advisory referendum; as its date approached, however, most parties rushed to declare that they would abide by the result.

Supporters of the Constitution saw it as a strengthening of human rights and democracy, while opponents feared the creation of a European 'superstate' in which the Netherlands would no longer have any say. Fear played a major role in these decisions: fear of loss of national identity and culture; of the Dutch language and influence; of social security and employment. And fear of Turkey's admission to the EU.[363] Most of these themes had nothing at all to do with the European Constitution, but they were linked to the debate by right-wing and left-wing populists.

In addition to fear, anger also played a major role. There was resentment about the high financial contribution which he Netherlands made to the EU. And about the fact that Germany and France violated EU financial rules but – contrary to EU regulations – were not fined for doing so. It did not get any better when people heard the director of the Dutch National Bank say that the guilder had been undervalued at the time of the transition to the euro in 2002: the Dutch had been robbed!

In short, the people were angry with Europe. And also at the Dutch government, which was held partly responsible. Since the government was in favor of the Constitution, many angry citizens were against it.

Suddenly, the feelings of fear and anger were stirred up violently in November 2004, when filmmaker Theo van Gogh was murdered by a fundamentalist Muslim, evoking memories of the murder of the populist politician Pim Fortuyn in 2002.

A wave of arson attacks on churches and mosques followed. At this point Dutch society was tight as a string.

In those circumstances, the complicated constitutional texts offered an ideal opportunity for manipulation by anti-globalists, anti-immigration and anti-EU parties and anti-establishment grumblers,[364] in short: anyone who was 'against' anything.

Only a month and a half before the referendum, the referendum committee published a brochure explaining the Constitution. The brochure was too late and too complicated[365] and was therefore heavily criticized.

On 1 June 2005, the people of the Netherlands had their say. The result was: 24.4% voted for the Constitution, 38.9% voted against, 36.7% did not vote.[366]

The House of Representatives concluded that the ratification could not go ahead, after which the government withdrew the proposal and the Netherlands therefore blocked the European Constitution. For the whole of Europe.

In France, the same thing happened on 29 May, so the Dutch result no longer really mattered: a veto by just one country was enough to kill the European Constitution.

Afterwards, experts analyzed the fiasco and drew conclusions. Referendums like that on the EU Constitution tend to fail, for a number of reasons, they argue. For example, the subject is too complex to understand, except for the experts. It is also multidimensional: it deals with a large number of issues that actually each deserve a separate decision process, but are compressed into one question. This oversimplification is compounded by the fact that no nuanced answer is possible to that one question, but only 'yes' or 'no'.

This creates enormous scope for populist manipulation that reinforces existing emotions or even evokes new ones. Since this is especially successful with negative, partly irrational emotions ('gut feelings'), the manipulation particularly strengthens the 'no' vote. "A referendum is more suitable for pulling down than to build up Constitutions", writes one researcher,[367] who suspects that the U.S. Constitution would never have been adopted in 1788 if a popular referendum had been required in each of the thirteen states.

These reflections by experts show that referendums are dangerous tools. This was demonstrated twice in 2016. First in the Netherlands, when the people were allowed to decide by referendum on an Association Treaty between the EU and Ukraine. The result: 12.3% accepted the treaty, 19.7% rejected it, 67.8% did not vote. Thus, 2.5 million Dutch naysayers blocked a treaty for 44 million Ukrainians and half a billion EU citizens that would not – as the populists lied – have automatically led to Ukrainian EU membership.

Two and a half months later, Britons voted in a referendum on Brexit (that is, the UK leaving the EU). For leaving the EU, 37.5% voted, while 34.7% wanted to stay; 27.8% did not vote. A spate of reports suggested that the referendum had been heavily manipulated by British populists, with help from Russia, including through tens of thousands of fake accounts on Facebook and Twitter. "How our democracy was hijacked", the respectable British newspaper *The Guardian* headlined.[368]

Referendums are not good decision-making procedures. They oversimplify complex issues. They have a polarizing effect; they pit sections of the population against each other; and they create fear, deceit, distrust and anger. Good decision-making procedures do the opposite: they lead to consensus, or at least to conclusions that are widely supported in society.

20.2 Inclusivity

Back now to the assessments I have carried out. Among the regular participants in these is at least the management, that is, one or more representatives of those who are authorized to make decisions about the subjects dealt with in the assessment. On one occasion, this requirement resulted in an incident. As agreed, I arrived early in the morning at the location where the assessment had been planned. There, the director came immediately to me and informed me that "unfortunately, I cannot participate, as something unexpected has happened that needs to be addressed right away." To which I replied that, regrettably, I was then forced to cancel the assessment and go home, as the participation of the director was indispensable. (The assessment went ahead. The director made the best of a bad job, let others solve the problems and participated.)

Those who have the authority to make decisions: that could be department heads, directors, deans of faculties, boards of directors, CEOs. Or governments. See what happens when a government curtails choice in advance and does not participate itself. And shudder.

The broad societal discussion on nuclear energy

For years, the use of nuclear energy had been under heavy fire in the Netherlands. The country possessed two nuclear power plants: Borssele and Dodewaard.

The battle between supporters and opponents took a new turn when a serious accident occurred at the Harrisburg nuclear power plant in the United States in 1977. The following year, the Dutch government decided that a *Broad Societal Discussion* (BSD: '*Brede Maatschappelijke Discussie*', BMD) would take place about the deployment of nuclear energy in the Netherlands. This was new: such a form of active involvement of the Dutch civil society had never happened before.

However, the plan received strong criticism when the government opened the discussion in a rather partisan manner with a 1,200-page report claiming that nuclear energy was 'inevitable'. That claim was dropped a year later, but the tone was set, and it was reinforced after the government announced in 1980 that the two existing nuclear power plants were not to be questioned in the BSD. That decision was reversed by the House of Representatives.[369]

It was not until 1981 that a Steering Committee was set up to lead the BSD. Its remit was "*to do justice to the way in which opinions are formed among all sections of the population and to allow these to be reflected in its advice to government and parliament*". Thus, a deliberate distance was created in advance between two parties: population versus politics.

Even before the Steering Committee started, the Deputy Prime Minister proposed not to wait for the BSD and to decide immediately to build new nuclear power plants, although a poll showed that 85% of the population was opposed to these plants. He was supported by the Prime Minister, who said, "I myself am already convinced of the necessity of nuclear energy. But in the Netherlands, a BSD is first needed to instill this in the people."[370]

These were the rather bizarre circumstances under which the BSD began. The discussion consisted of two phases.[371]

> Between September 1981 and October 1982, the *information phase* took place, during which the population was called to participate through a series of advertisements in all major newspapers. In addition, 160 institutions and groups were directly invited. Numerous organizations and individual citizens answered the call. The Steering Committee received 3952 responses, including reports, statements, pamphlets, and signature lists.

The information phase was completed with the creation of four scenarios that became the focus of the second round, the discussion phase. Around that time, the (now next) government reiterated that, whatever the outcome, the existing nuclear power plants would continue to operate no matter what.

> In the *discussion phase* of the BSD, there were approximately 2,000 public meetings with experts. Municipalities, provinces, civil society organizations and hundreds of schools were actively involved. An information newspaper on the BSD was distributed with a circulation of 1.1 million. The discussion cost a total of 27 million guilders.
>
> The results in 1984: a large majority of the population turned out to be opposed to nuclear energy. Almost 80% did not want any new nuclear power stations; 57% even thought that the two existing nuclear power stations should be closed down immediately.[372]

Later in the same year (1984), a memo from the Ministry of Economic Affairs leaked out, showing that the government was working on plans to build two to four new nuclear power plants, the first of which was to be commissioned in 1995. In 1985, the House of Representatives agreed to this.[373] Location procedures had already begun when the nuclear power plant at Chernobyl (Ukraine) exploded on 26 April 1986, a Category 7 accident, the worst possible nuclear disaster. In the

Netherlands, the cows had to be stabled because of the radioactive fallout; home-grown vegetables could not be consumed. That was the end of the government's plans for new nuclear power plants. At least until about 2010, when the discussion was revived by the government and the nuclear energy lobby. That process, too, ended abruptly with a Category 7 nuclear disaster, when the Fukushima nuclear power plant in Japan caught fire on 11 March 2011.

In 2020, the nuclear lobby restarted. (For details, see Appendix 10.)

Many declared afterwards that the BSD had failed. But that conclusion is too one-sided. The discussion itself, despite all the opposition, went extremely well. The involvement of tens of thousands of citizens and of media, education, companies, experts and civil society organizations was great. The process was reasonably structured. The conclusions were well founded and gave a realistic picture of the views of the well-informed participants.

What did go gravely wrong, however, was the way in which politicians, especially the government, deceived the population with the BSD. The course of events clearly betrays that the government was not at all keen on a serious public discussion, had no intention of listening to the results, and wanted only one thing: to push through nuclear power. As a result, the BSD was not democracy but demagogy, and thus – as a democratic experiment – it had indeed failed.

Problem ownership

The example of the BSD shows why it is so important that the decision-makers, the competent authorities, participate in the decision-making process. That they are *genuinely* and *curiously* involved; that is, not with a hidden mission. Those who don't participate, have no input into opinion formation, do not listen to the views and emotions of others, and therefore don't become co-owners of the chosen solution. This creates distance or even distrust instead of togetherness. After that, it is all too easy to sit back and reject the solution. Consensus is a *process*, not a *decision*.

Participatory democracy

Churchill was wrong: a BSD is a much more democratic decision-making process than majority decisions by voting. But there are conditions.

- **Inclusivity conditions**

> Ensure that everyone participates who is (or should be) a co-owner of the issue.

To participate can also mean: to be represented.

'Everyone' means: every individual and group of people, including associations, interest groups and demonstrations; villages, neighborhoods and cities; companies, foundations and organizations; institutions such as the police, healthcare, education, centers of excellence.

Since politics definitely belongs to these co-owners, politics must actively participate, otherwise the procedure has no chance. Politics, that is: the political parties; the legislature (parliament, provincial states, municipal council); the administrators (president, government, provincial executive, mayor & aldermen). Equally indispensable is the active participation of the 'silent power', the civil servants.

Ensure that every input is taken seriously.

Don't judge that another's input is insignificant, or bland, or irrelevant. Each participant determines that for themself. 'Small' local issues dominate the lives of many[374]*: accept that and do something with it.*

Ensure an appropriate representation of those who cannot represent themselves.

Represent the voiceless. The children. The mentally challenged. The distant and unattainable. The despondent ones.

Represent the animals. The natural world. The future generations. Maybe even: the ancestors.

Ensure that only those with an interest in the issue participate, not uninterested parties.

Do not call for everyone to participate. Such an appeal to the uninterested pollutes the process and is anti-democratic.

Ensure that the conclusion is also the decision.

Once the discussion has reached a conclusion, all participants, including politicians, must accept it and turn it into a formal decision.

- **Truthfulness conditions**

Ensure rootedness in the facts.

No myths, please. Let science and education speak. 'Alternative facts' are lies. Respect expertise.

Ensure a balance between facts, opinions and emotions.

Discover at which point facts are relevant and at which point beliefs, traditions, emotions, dreams, ideals.

Ensure a discussion, not a debate.

The goal is: to find the best solution, not to win. Let the discussion be a quest that is only successful if everyone wins.

Ensure that the powerless speak (first) and the powerful listen to them.

This is not about finance. The discussion succeeds if the rich and powerful also experience that they win, even in the case of redistribution of power and money.

Ensure that the discussion is not manipulated.

I admit that I don't know a solution to the problem of mass manipulation, for example, through online social media. Who does?

- **Implementation conditions**

Ensure excellent discussion leadership.

Executives, administrators, governments: they are not the moderators of the discussion but participants with a special role: they are going to implement the conclusions.

Ensure that the process is allowed to mature; don't force it.

Creating consensus and support takes time. The discussion is not an event but a process: in complex issues even a never-ending process.

Only draw conclusions when it is clear to everyone that the time is ripe. Check whether the conclusions are indeed supported.

Ensure the use of a wide variety of methods and techniques.

Leverage the new tools, such as social media, online meetings, big data. But protect privacy.

Give space to every style of thinking and expression: science, philosophy, religion, literature, science fiction, visual arts, cartoons, cabaret, poetry, song, dance, play.

> Ensure well thought-out, openly visible direction.

Create a roadmap with milestones in advance. First create support for the route, then implement it. Adjust it flexibly, whenever necessary, then again ensure support.

You know what participation is *not*? Let me give you an example. During an assessment, I often ask, "Do you have a mission statement?"

Usually the director then proudly says, "Sure we do!"

I: "How did that come about: were the employees involved in that?"
DIRECTOR: "Yes of course!"
ME: "Aha. And how did that go?"
EMPLOYEE: "He made a draft text. We could respond to that."

Well, then I know enough. That's *not* participation. The text is the property of the director, not the staff. The staff is then allowed to make some comments. If they're lucky, the text will be amended here and there. But the employees will never see it as 'their' text again, and the mission statement has become meaningless: a missed opportunity. Fortunately, there are also companies and institutions that do take participation seriously.

Something similar applies to democracy with the help of broad societal discussions. Here, too, it is not a question of creating societal support afterwards for political decisions that have already been taken, but of active participation from the start. In other words, real *participatory democracy*.

Transdisciplinary science

The purpose of this book goes even further: it is about society's participation in science. When all kinds of sciences transcend the boundaries of their fields, they work interdisciplinary. When society is also actively involved, it is *transdisciplinary science*.[375] This goes much further than one-way traffic, in which science informs or educates society.[376] It also goes beyond one-way traffic in the other direction, in which society places limits on scientific research. Transdisciplinary omniconomic cooperation means that society and science, in permanent mutual consultation, give direction to scientific research, with science's mission: build the world in which we, based on broad support, want to live in. Omniconomics thus goes a step further than the usual transdisciplinary sustainability science,[377] which largely focuses on glued-on sustainability. The task is: bring the global society out of puberty; lead us to a stable and healthy adult humanity. Make the world intrinsically sustainable.

The Broad Holloway

During the assessments I conducted in companies and educational institutions, I always managed to reach consensus, which often involved complex decisions on difficult, fraught issues. But those organizations always had a limited size, so it was relatively easy.

Many towns and villages are experimenting a great deal with civic forums[378] and citizens' platforms.[379] On that scale, 100% consensus may perhaps be too high an ambition; but broad support is achievable, even at the level of an entire country. The BSD on nuclear energy is an excellent example: if such a BSD is implemented next time *with* rather than *opposed by* politicians, truly democratic decisions can be made.

My proposal is:

We conduct a well-directed international Broad Societal Discussion, in which science, citizens, business, civil society organizations and politics together create a broad basis of support with respect to the world in which we want to live.

The omniconomic task for science is then, to discover in a transdisciplinary way how that world can be realized. This involves intensive use of experiments in virtual realities. The possibilities and impossibilities discovered in the process are fed back into the BSD, creating an iterative, never-ending process.

The envisioned world is dynamic and as stable as possible, so that we move into the future along a Broad Holloway, as in the complex system consisting of nature, humanity and economy (planet, people & profit) the attractors are directed towards the middle of the Holloway instead of the edges.

The dance

You are standing at the front, by the white smartboard. You are the president of a civil society organization; across from you sits a group of about thirty experts. In a ring around you are hundreds of interested people.

Through the internet connection, thousands more people are watching, including members of committees from more than twenty cities with whom you have prepared this. They will soon be providing additions and comments.

You explain the ideals of your foundation with all your body. You run back and forth drawing lines on the white board, the screen fills with colorful lines and symbols. You make broad gestures with your hands: that line goes from there to there, so that it...

At the same time, you explain in words what the people in your teams dream about and why. You think out loud. You tell a narrative, your voice going up and down, sometimes whispering to convey the miraculous beauty of your ideals. You let the biologists, the planners, the psychologists, the business experts, yes even the physicist complete your sentences, ask questions to the politicians and journalists, get answers. You perform a dance with the group.

As you do so, you look carefully at the experts, at each of them. You listen to what they say; while narrating, you pay attention to their body signals.

The miracle happens. You see their eyes light up and you know: they ***understand****.*

Sizzling, the conversation about your proposals begins.

20.3 Personal message to the reader

Esteemed reader, at the time of writing I have been working on sustainability for over thirty years, with all my passion and inspiration. I have designed academic courses dedicated to it. I have coached university and college boards and principals, developed strategy plans with them, and instructed lecturer teams in reworking curricula toward intrinsic sustainability. I wrote textbooks, management methods, and my PhD thesis.

That was wonderful work, but it all falls short of this book, which has become my *magnum opus*. I have faith that the proposal with which I just closed this chapter can make a far greater contribution to a better world and a more beautiful society than all my previous work put together.

And now I have a request for you. If you find the proposal as valuable as I do, and want to give it a chance: tell others about it passionately. Help spread the ideas, because I can't do it on my own.

See if you can manage to get two others to read this book. If they do the same, then a development is started that can make the plan a reality.

And there is more that you can do concretely.

If you want to take social initiatives yourself to get the proposed BSD going – locally, nationally, internationally: contact me (https://niko.roorda.nu/contact). Involve existing civil society organizations, or set up new ones. Involve your business, if you can. Involve your neighbors, your friends, your family. Let it sizzle.

If you want to take scientific initiatives to explore one of the themes of the plan in more detail: contact me, do research, put graduate students, PhD students or departments to work, improve or add to my appendices, write new appendices.

If you can develop appropriate software (virtual reality, Artificial Intelligence, evolving A-life) for the purpose of *survival of the sustainablest*: contact me. Also contact scientists.

If you can play a coordinating, a funding or a media role in the complex processes I write about: contact me.

In this way, with many in a transdisciplinary cooperation, together we build a healthy and resilient omniconomic science and society.

I salute you and wish you and me an intrinsically sustainable future.

Recommended further reading (for details, consult the online Bibliography)

Journal article: Roorda, Niko (2018): *Future-Focused Entrepreneurship Assessment (FFEA).*

Journal article: Hobolt, Sara Binzer & Sylvain Brouard (2011): Contesting the European Union? Why the Dutch and the French Rejected the European Constitution.

Journal article: Podolnjak, Robert (2006): 'The Assent of the People is Not Necessary to the Formation of a Confederation': Notes on the Failure of the European Constitutional Referendums.

ACKNOWLEDGMENTS, CREDITS, LITERATURE, APPENDICES

The author thanks:

The publishers: Annabelle Harris and Caroline Church, Routledge.

My wife Marjo van Giersbergen-Roorda, who was always the first to provide critical commentary on the texts and constantly supported and encouraged the author.

The second ring of positive critics: Christiaan Roorda, Anjo Roorda and Juliëtte Schaafsma.

The think tank of young creatives: Tom Ballering, Stefan Berrelkamp, Gé Cuypers, Mirjam van Doorn, Thom Hamer, Guido Houben, Marlou Jager, Daniëlle Janssen, Robin Kuijpers, Marieke Peterson, Lennart Quispel, Michiel Roorda, Sophia Scheper, Wessel Verhaar, Julia van de Warenburg, Vivian Yip.

The team of the Dutch online journalist forum *Follow the Money*: Arne van der Wal, Jan-Willem Sanders, Eric Smit, Maartje Laterveer, Luuk van der Sterren. And in the Forum discussion: Jonathan Barth, Michiel Dijkman, Michel Fleur, Anton van de Haar, Jean Klijnen, Lydia Lembeck, Pieter van Loon, Annemiek van Moorst, Nick, Co Pater, Milco Schmidt, Gerrit Zeilemaker and many others.

For his coaching: Pim van Tol.

Platform DSE and Our New Economy Foundation: Esther Somers, thanks to whom I started the Omniconomics project.

Critic and thinking partner of the earliest texts, more than thirty years ago: Jann Mens.

Brainwash, Human: Tjebbe Venema. Noordhoff: Vincent Diks. And further: Sjoerd van Bekkum, Frans Doorman, Ruben Graafland, Jan Jonker, Kees Klomp, Willem Schramade, Kees Uittenhout.

Credits

All images were created by the author, with the exception of:

Figure 2.2: Scan.
Figure 3.5: Background photo: NASA Earth Observatory.
Figure 4.6: Romain Bréget on Wikimedia: Chemin creux du coté de La Meauffe.
Figure 5.1: Tycho Brahe, 1602.
Figure 6.2: DTR on Wikimedia.
Figure 6.3: Johannes Kepler: Mysterium Cosmographicum (1596).
Figure 6.4: NASA/JPL, Viking 1 Orbiter (1976); NASA/JPL, Mars Reconnaissance Orbiter (2007).
Figure 9.1: Donaldytong on Wikipedia.
Figure 12.2: Background photo: NASA Goddard Space Flight Center.
Figure 14.1: John William Waterhouse: A Naiad or Hylas with a Nymph.
Figure 14.2: Sofia Vini on Wikimedia; NASA Earth Observatory.
Figure 14.3: National Institute of Allergy and Infectious Diseases (NIAID).
Figure 15.4: EHT Collaboration, emailed permission.
Figure 18.1: GerryShaw on Wikimedia; Mike Lee on Flickr.

Some figures are taken from Roorda's *Fundamentals of Sustainable Development*, courtesy of Routledge Publishers.

Appendices and Literature

The links to the Literature and the Appendices are, as announced in Chapter 1, at https://niko.roorda.nu/books/omniconomics.

NOTES

Chapter 2

1 ISDA (2016).
2 EC (2012).
3 Reuters (2013).
4 US News (2016).
5 Lin (2017).
6 Lucas (2003).
7 Bernanke (2004).
8 Cotis (2007).
9 See e.g. Fama & French (2012).

Chapter 3

10 Elkington (1998).
11 Roorda & Rachelson (2018).
12 ILO (2017).
13 Lino et al (2017).
14 Asdal (2021).
15 In my English-language book, with co-author Anouchka Rachelson: *The Seven Competences of the Sustainable Professional* (Routledge, 2018), and in my Dutch book *De Zeven Competenties van de Duurzame Professional* (Garant, 2015). See: https://niko.roorda.nu/books.
16 Good original literature sources are: DfD: Boothroyd & Alting (1992). LCA: Heijungs et al (1992). Integrated chain management: Cramer (1996). C2C: McDonough & Braungart (2002). Circular economy: Pearce & Turner (1989). Doughnut economy: Raworth (2017). An overview of the entire development, from 'Lansink's Ladder' to 'doughnut', is offered by the designer of the ladder, Lansink (2017).

Chapter 4

17 Roorda (2008).
18 Van den Bergh (1960).
19 Clarke (1962/2000).
20 Fukuyama (1992).

Chapter 5

21 Brahe (1602).
22 ἐνέργεια ἀσώματος κινητική. (Philoponus, 517 A.D.).
23 Inclinatio. (Avicenna, 1020 A.D.).
24 Shawq. (Al-Bitruji, ca. 1200 A.D.).
25 Newton (1713).
26 Stahl (1718).
27 O.a. Lavoisier (1784).
28 Reichenbach (1845).
29 Blavatsky (1892).
30 Wexler (director, 2011).
31 Şengör (1982).
32 Libet (1999).
33 Freud (1923).
34 UNEP (2011).
35 Henckens (2016).
36 Listner (2011).
37 USPTO (1998).
38 SCOTUS (2013).
39 Cole (2015).
40 Smith (1887).
41 Clark (1985); Kaiser (1987); Nooij (2007).
42 Zussy (1993).
43 Perry (director, 1971).

Chapter 6

44 Kuhn (1962).
45 Kuhn (1970).
46 Hulspas & Nienhuys (1997).
47 Kölreuter (1761, 1763, 1764, 1766).
48 Lamarck (1809).
49 Babbage (1837).
50 Chambers (1844).
51 Lyell (1832).
52 Cuvier (1827).
53 Behe (1996); Dembski (1998).
54 Dawkins (1996); Miller (1999). Evolution biologist Richard Dawkins dedicates in his chapter 5 ample attention to a famous issue: the genesis of the eye.
55 Such as: Chang (2014), Snowdon & Vane (2005), LK (2014), Samuelson & Northaus (1985).
56 Durden (2014), Chang (2014). With additions through a forum discussion by Jean Klijnen, Annemiek van Moorst, Anton van de Haar, Gerrit Zeilemaker.
57 Fox (2013).
58 Grossinger (1986); Hoagland, Richard (2002). Vinous comments were provided by Sagan (1995) and McDaniel & Paxson (1998).
59 Data acquired from GGDC (2016).
60 See e.g. Ledenyov & Ledenyov (2013).
61 Burns and Mitchell (1946).
62 NBER (2010).
63 Korotayev & Tsirei (2010); Škare & Stjepanović (2015).
64 Solomou (1986); Bieshaar and Kleinknecht (1986); Rosser (2013).
65 Smith (2016); see also: Smith (2013).
66 Vigen (2015).

Chapter 7

67 Kuznets (1955).
68 Deininger & Squire (1998).
69 Hertz (2001); Notzon et al. (1998); Stuckler et al. (2009); Milanovic (2011).
70 Piketty (2013).
71 In: Rankin (2013).
72 See e.g. Houben (2017).
73 Gossen (1854).
74 Among others by Daniel Kahneman (2011).
75 Ricardo (1815; 1817).
76 Findlay (1987); Harrigan (2003).
77 MacDonald and Markusen (1985).
78 Samuelson (2004); Schumacher (2013); Shaikh (2016).
79 Coase (1960).
80 See e.g. Rifkin (2004).
81 Marx & Engels (1848).
82 Orrell (edition 2017).
83 Nadal (2015).
84 Reis (2018), p. 140.
85 De Grawe (2009); Skidelsky (2009).
86 Comte (1830, 1853).
87 Tinbergen (1962).
88 Baier & Bergstrand (2009).
89 Descriptions and updates are to be found in Arrow & Debreu (1954) and McKenzie (1954 and 1959).
90 Stiglitz (2018), p. 76.
91 Blanchard (2018), pp. 45–47.
92 Hardt (2021).
93 Pashler & Wagenmakers (2012); Schooler (2014); Smith (2017).
94 Camerer et al (2016); Bohannon (2016).
95 McCullough, McGeary, Harrison (2006).
96 Blanchard (2018).
97 Stiglitz & Walsh (2006), pp. 6–7.

Chapter 8

98 Gilboa (2015).
99 O'Neill & Handal (2021).
100 Unfortunately, no source; and that is precisely the point.
101 Sources: Kremer (1993) and UN DESA (2017), middle scenario.
102 Sources: Bradford DeLong (1998) and IMF (2018).
103 Sources: Bradford DeLong (1998) and IMF (2018).
104 Winter & Leenaarts (2015).
105 Beunder (2015), referencing Heilbroner (1953).
106 Ayres (2008); Jackson (2009).
107 Kalshoven (2008).
108 Lomborg (2001, 2004).
109 Mulder & Koster (2008).
110 Roorda (2021), Chapter 5.
111 In an interview with Mulder & Koster (2008), pp. 69–74.
112 Dittrich et al (2017).
113 Harrod (1965), p. 77.
114 Among others: Kennedy (1995); Jackson (2009); D'Alisa et al (eds., 2014); Daly (2014).
115 Meadows et al. (1972).

116 Solow (1973). See also Georgescu-Roegen (1971) and Erreygers (2009).
117 Among others: Voet et al (2005); Dittrich et al (2017).
118 Hickel & Kallis (2019).
119 Lent (2017).
120 Daly (2014).
121 Sources: Financial Times (2017), Reuters (2017).
122 Source: Van Dale groot Woordenboek van de Nederlandse taal, 15th edition. (Boon & Hendrickx (eds.), 2015).
123 Source: Collins English Dictionary, 12th edition, 2014.
124 Source: Aletta André in Dagblad Trouw, 27 September 2017.
125 Source: Leonie Hosselet in De Volkskrant, 4 August 2017.
126 Source: Tony Sheldon in British Medical Journal, 10 May 2018.
127 Source: Michiel van der Geest & Ellen de Visser in De Volkskrant, 16 June 2018.
128 Source: Michiel van der Geest & Ellen de Visser in De Volkskrant, 16 June 2018.
129 Source: Ellen de Visser in De Volkskrant, 28 May 2016.
130 Source: Margreet Vermeulen in De Volkskrant, 18 September 2015.
131 Hordijk (2019); Van Laarhoven (2019).
132 Fantasy of the author; but you knew that, didn't you?
133 Fernandez & Klinge (2020).
134 Baker (2004); Tieleman & von Piekartz (2020).

Chapter 9

135 Mazák et al. (2011).
136 Goodall (2010).
137 Mitani (2010).
138 Merritt (2010).
139 Plato (ca. 380 B.C.); Aristoteles (ca. 350 BC); Hobbes (1642); Hume (1739); Karl Marx (1862).
140 Locke (1689).
141 Hublin et al. (2017).
142 Dawkins & Wong (2016).
143 Hershkovitz et al. (2018).
144 Pauketat (ed., 2012).
145 Hunt & Lipo (2006).
146 Diamond (1997); Harari (2014).
147 Rousseau (1755).
148 Lescarbot (1609).
149 VPRO (2017).
150 Reumer (2017).
151 Bowman (1995).
152 Beach et al. (2019).
153 Harari (2014).
154 Higham et al. (2014); Devièse et al (2021).
155 Dorey (2015).
156 IUCN (2021).
157 Cavalieri & Singer (eds., 1993).
158 Precht (2016).
159 Van de Voorde (ed., 2021); Meijer (2017).
160 Lovelock (1979); e.g. in *Theogony* by Hesiod (8th or 7th century BC).
161 See e.g. Latour (2017).
162 Yampolskiy (2021).
163 Cottrell et al. (2014).
164 Koplin & Wilkinson (2019). See also Devolder et al. (2020) and Piotrowska (2021).

165 Knoppers & Greely (2019).
166 Bloomberg (2017).
167 Guardian (2021).
168 Oxfam (2016).
169 Oxfam (2019).
170 Roorda (2020).
171 Oxfam (2021).
172 World Bank (2020).
173 Kohler (2017).
174 Brandmeir et al (2017).
175 Piketty (2013; English edition 2014; Dutch 2016).
176 Reuters (2021).
177 Rosling (2018).
178 Pinker (2018).
179 Griffiths & Gallón (2018).
180 Survival International (2019).
181 Other sources: Batchelor (1905); Free.fr (2018); Schorer (2016); Omniglot (2018); Naish (2014); Wikipedia List of languages by time of extinction (Wikipedia, 2018); Wikipedia List of numbers in various languages (Wikipedia, 2019).
182 Van de Weijer (2018), De Volkskrant.
183 Klein (1999).
184 Martijn & Tokmetzis (2018).
185 House of Commons (2019).
186 ILO (2017).
187 Semu-Banda (2008); Roorda (2020); Liew (2017).
188 Walmart (2019).
189 Brandwatch (2019).
190 Knapton (2016).
191 Hertz (2020).

Chapter 10

192 Tainter (1988).
193 E.g. Greer (2005).
194 Oreskes & Conway (2014).
195 Meadows et al. (1972). In Dutch, the book was given the incorrect subtitle 'Rapport *van* de Club van Rome'.
196 Passell et al. (1972).
197 Bailey (1989); Bailey (1994).
198 Bardi (2008); see also Bardi (2017).
199 Nørgård et al. (2010).
200 Meadows et al. (2004).
201 https://insightmaker.com/insight/1954/The-World3-Model-Classic-World-Simulation.
202 Lawton & May (1995).
203 Leakey & Lewin (1995); Kolbert (2014).
204 IPBES (2019).
205 Plumptre et al. (2021).
206 Quammen (2013, 2020).
207 Altieri & Nicholls (2020).
208 Ahmed et al. (2019); Vaughan (2021).
209 IMF (2021).
210 NOAA (2019).
211 NOAA (2021).

212 NOAA (2017).
213 Rahmstorf et al. (2015).
214 Hansen & Sato (2015). IPCC (2021).
215 Shepherd et al. (2018).
216 Rosier et al. (2021).
217 Breitburg et al. (2018).
218 Raad Volksgezondheid & Samenleving (2021).
219 Kindleberger & Aliber (2005).
220 Shaheen & Dehghan (2015).
221 Kang & Eltahir (2018).
222 Carrington (2018).
223 Im et al. (2017); Carrington (2018).
224 Diamond (2005).
225 Van Bavel (2016).
226 Stiglitz (2012); Walter Scheidel (2017).
227 Greer (2005).
228 Turner (2014).
229 Ehrlich (1968); Ehrlich & Ehrlich (2009).
230 Drexler (1986); Vinge (1993); Kurzweil (2005).
231 Reinhart & Rogoff (2009).

Chapter 11

232 Woodford (1999).
233 Mazzucato (2018).
234 Morgan (2001).
235 Ingrao (2018); Róna (2021).
236 Morgan (2012).
237 Focardi (2015).
238 Róna (2021).
239 Morgan (2001).
240 Reagan (1967).
241 Morgan (2001).
242 Keen (2011); Krugman (2009); Taleb (2007 / 2010).
243 Gilboa (2015).
244 Petram (2011).
245 Buser & McConnell (2014).
246 Williams (2013).
247 Gaddis (1986).
248 Clauset (2018).
249 Fisher (2017).
250 Christiano et al. (2017).
251 Thaler (2015).
252 Hawken (2010); Krugman (2010); Beeks & Ziko (2018).
253 The term was introduced by H. Eugene Stanley in 1995. See: Stanley (2013).
254 See e.g. Hausmann & Hidalgo et al (2011).
255 Harari (2014), Chapter 12.

Chapter 12

256 WCED (1987).
257 UNCED (1992).
258 UN (2000).
259 UN DESA (2015).

260 Earth Charter Initiative (2000).
261 Geodata (2016); NASA OzoneWatch (2018).

Chapter 13

262 Kohnstamm (1987). Jean Piaget speaks of *thinking structures*, in French: *schèmes*: Piaget & Inhelder (1966).
263 Piaget (1970).
264 Descartes (1644).
265 Hume (1738).
266 Precht (2007).
267 Le Verrier (1858); Adams (1846).
268 Duhem (1906), followed half a century later by Popper (1959; 1963). For an overview, see Maxwell (2014).
269 Gunster (2010).
270 European Commission (2010), recommendation 6.
271 Wang et al. (2013).
272 De Waard (2019).

Chapter 14

273 Shermer (1997).
274 Durkheim (1912); Anderson (2005, p. 3).
275 De Brès (1561).
276 Aristoteles (ca. 350 BC): Categorias.
277 Majjhima Nikāya (3rd century BC to 2nd century AD).
278 Kant (1781).
279 See e.g. Malinowski (1948); Anderson (2005).
280 Sarfati (2011).
281 NAS (1999).
282 Bacon (1605); Çimen (2016).
283 Rowland (1883), pp. 243–244 (shortened).
284 See: Neurath (1938–1970); Polanyi (1962).
285 NSF (1953), p. 38).
286 Bullock & Trombley (eds., 1999).
287 Popper (1934; 1959).
288 Popper (1989).
289 E.g. Gasperinia & Veneziano (2003); Bojowald (2007).
290 E.g. Penrose (2006); Salah et al. (2017); Das et al. (2018).
291 Hawking & Hertog (2018).
292 Goff, Philip (2019).
293 Sha & Xiu (2018).
294 Edwards (2001); Ford (2003).
295 Szilárd & Einstein (1939). See also Hewlett & Anderson (1962).
296 Bodenheimer (2000).
297 Robbins (1932).
298 West (2018).

Chapter 15

299 Kunckel (1679); see Szabadvary (1985).
300 Home (1756).
301 Hooke (1665).

302 Hollander, de (2018b), referring respectively to Ogawa et al. (1990); IHGSC (2004); Bishop (2006); van der Zwaag et al. (2016); McCoy et al. (2017); Nishimoto et al. (2011); Manyika et al. (2011).
303 Toulmin (1953).
304 Manglik et al. (2006); Kieffer (2016); Badshah et al. (2018).
305 Mendeleev (1869).
306 Kaji (2002); Kak (2004).
307 Mendeleev (1902).
308 See e.g. Novella (2010); Gorski (2010).
309 Morgan (2001).
310 Muchowska et al. (2019).
311 Al-Khalili & McFadden (2014).
312 Jogalekar (2013).
313 FAO (1997).
314 Hall & Rao (1999); Zijffers (2009); Blankenship et al. (2011).
315 Lakatos (1978); see also Lakatos (1970).
316 Ijjas et al. (2017).
317 Schwarzschild (1916).
318 Verlinde (2011).
319 Feyerabend (2004, p. 72).
320 Kuhn (1962).
321 Ptolemaeus (ca. 150).
322 Copernicus (1543).

Chapter 17

323 Simon (1962, p. 267).
324 Roorda (2020).
325 Prigogine (1996).
326 Lewin (1992).
327 Strogatz (2003).
328 Frans de Waal (2019).
329 Bertuglia & Vaio (2005).
330 Modlmeier et al. (2019).
331 Augustin (2004).
332 Steffen et al. (2018).
333 Petit et al. (2001).
334 Petit et al. (2001), Grant et al. (2014).
335 Scheffer (2019).
336 Kirman (2011), Rosser (1999, 2013).
337 Blume & Durlauf (eds., 2006).
338 Rosser (2001).
339 The economic ideas of Keynes: Citera (2016, 2017). DSGE: Fagiolo & Roventini (2017). Arrow-Debreu: Gosselin et al (2018). Unfortunately, the new approach does not differ substantially from the old one, as equilibrium points are still sought based on abstract mathematical models that are not realistic: see, for example, Moran & Bouchaud (2019). Complexity is applied to the design of corporate and public policies: Sterman (2000). Finally, there are the attempts to arrive at an Econophysics, a physics-like approach to economics in a complex environment: Stanley (2013), Bouchaud (2019). A fine overview has been published in the form of an atlas: Hausmann & Hidalgo et al (2014).

Chapter 18

340 Menichetti, Dall'Asta, Bianconi (2016).
341 Moliterno & Mahony (2011); Zappa & Robins (2016).
342 Abshoff et al. (2014).
343 Battiston et al. (2018).
344 Hutchinson (2018).
345 Heisenberg (1927).
346 Hobson (2013).
347 Pointon (1967).
348 Al-Khalili & McFadden (2014). Quote taken from Black Swan edition (2015, pp. 45–46).
349 Anderson (1972).
350 Durston & Baggerman (2017).

Chapter 19

351 Turing (1950).
352 Silver et al. (2017); Singh (2017); Okun & Jackson (2017); Gibney (2017).
353 Silver et al. (2018).
354 Moravec (1988).
355 Iten & Metger et al. (2019).
356 The AlphaStar team (TAt) (2019).
357 Waarlo (2019).
358 Witkowski (2021).

Chapter 20

359 Plato (ca. 380 BC): Politeia.
360 Churchill (1947).
361 Roorda (2008b; 2009; 2012).
362 Roorda (2018a; 2018b).
363 Hobolt & Brouard (2011).
364 Siedentop (2005).
365 Baden & de Vreese (2008).
366 Europa Nu (n.d.).
367 Podolnjak (2006).
368 Guardian, The (2017).
369 Stichting Laka (n.d.).
370 Trouw, 9 January 1981, quoted in Gabriëls (2001, p. 92). The Prime Minister in question was Dries van Agt, the said deputy prime minister was Hans Wiegel. See also van Vugt (2004, p. 46).
371 Stuurgroep BMD (1983); Hagendijk & Terpstra (2004).
372 Stuurgroep BMD (1984); see also Mulder (2012).
373 van Vugt (2004).
374 Visser (2008). The "driving judge" shows this illustratively.
375 Lang et al. (2012).
376 van Est (2011).
377 Keitsch & Vermeulen (Eds., 2021).
378 Vliet, Peter van (2021).
379 Burgerplatform Zaanstad (2021).

INDEX

accountable 226
adding up graphs 67
A$eists 188
Agenda 21 158
Agenda 2030 159
aggregation levels 201
AI 253
alienation 110
A-life 258
Allergan 101
AlphaGo 254
AlphaStar 255
AlphaZero 254
alternative medicine 211
AMI 183
animal dignity 115
Antarctica 113
anthropodiversity 115
anti-globalists 271
appendices, download 3
a priori knowledge 193
Aristotle 43
arithmetic mean income 182
Arrow-Debreu Model 81
artificial intelligence 253
artificial life 258
Atlantic Gulf Stream 138
attractor 233
Austrian school 63
automatism 74
automotive industry 105
average income 182
Babbage, Charles 61
babies and toddlers 168
banks 12
Barbarian Invasions 140
basic income 264
BAU 133
behavioral economics 153
behaviorist school 63
Bernanke, Ben 16
Berzelius, Jöns Jakob 47
Bespoke Tranche Opportunities 14
Biafra 51
bias 197
Bible 181
bibliography online 4
Big Bang 196
Bildungstrieb 47, 231
biocapacity 100
biology 47
Black Death 139
black holes 174
Blanchard, Olivier 82
Blessings of Capitalism Myth 77
Brahe, Tycho 42
brake energy 178
Braungart, Michael 30
BRCA1 54
breast cancer 54
Brexit 86, 271
Broad Holloway 40, 278
broad societal discussion 272
Brundtland Report 158

BSD on nuclear energy 272
Buddha 172
Burj Khalifa 117
Business as Usual 133
butterfly effect 230, 247
buying time 157

canon 212, 217, 222
capitalism 25, 78
causality 196
CDS 9
CFCs 162
chaotic systems 229
characteristics of omniconomics 221
check 190
chemical industry 105
chemistry 46
Chernobyl nuclear accident 273
chimeras 116
chimpanzees 111
chlorofluorocarbons 162
Churchill, Winston 268, 274
civic forums 278
Clarke, Arthur C. 37
climate change 136
climate cycle 239
closed system 223
Club of Rome 133
Coase theorem 77
Cogito ergo sum 172
coherence 222, 250
collapse 132
collateral damage 125
completeness 223, 250
complexity 223, 228
complexity economics 153
complex systems 229
computer simulations 234
Comte, Auguste 80
confirmation bias 197
conjuncture 67
connecting the dots 176
consensus 268
conservation laws 170
Conservation of Belief 192, 211, 219
Conservation of Energy 170
Conservation of Life (after Death) 170
Conservation of Money 186, 264
Conservation of Money Current 264
Copernican revolution 58, 218
Copernicus, Nicolaus 43, 218
corona crisis 139
Cosmic Cup 65, 210
Cotis, Jean-Philippe 16
covering roof 249
Covid-19 pandemic 136
cradle to cradle 30
creationists 61
Credit Default Swap 9
cross-disciplinary tools 250
cryptogram 177, 180
cuddle hormone 197
cultural extinction 123
cultural identity 122
curvature in space and time 174, 180
Cuvier, Georges 61
cycles 67

Darwin, Charles 61
deadly heat waves 139
Dead zones 138
debt cancellation 160
decoupling 99
Deep Blue 254; definition of; closed system 223; economics 88; fact 194; hypothesis 194; impetus word 46; intrinsic sustainability 33; law 194; omniconomics 35; profit 102; protoscience 59; religion 154; research area of omniconomics 204; scientific success 91; theory 194; Theory of Really Everything 181; triple P 19; understanding 176
dematerialization 98
democracy 160, 268
derivative 9
Descartes, René 172
design for disassembly 30
devastating attractor 233
Diamond, Jared 141
differential calculus 80
discovery 171
discrimination 126
disruptive 225
doughnut economy 30
download page 3
drug industry 103
DSGE 81
Dutch Credo 192
dynamic stability 224
Dynamic Stochastic General Equilibrium 81

eagercredulous 73, 153, 191, 221
Earth Charter 159
Earth Overshoot Day 100
Earth Summit 158
ecological footprint 100
ecological sustainability 19

econometrics 81
economics (definition) 88
economic growth myth 98
economic sustainability 20
econophysics 153
edge of chaos 231
efficient market myth 76
efficient markets hypothesis 16
Ehrlich, Paul 143
EIA 89
eightfold path to science 207
Einstein, Albert 174
Einsteinian revolution 218
Élan vital 47, 231
electromagnetism 214, 248
Elkington, John 19
emergence 231
emotional output 91
empirical 223
engagement criteria 225
environmental economics 153
Environmental Impact Assessment 89
epistemology 193
equilibrium 75
equilibrium attractors 233
ethics 193
Euro 86
European Constitution 270
European Green Deal 160
European Migration Period 140
European Union 86
evergreening 104
evolutionary process 260
exotic diseases 137
explanatory diversity 250
explanatory principle 196
exponential growth 94, 129
externality 24, 105
Exxon Valdez 8

face on Mars 67
fact 194
falsifiability 222
falsification 197
Fama, Eugene 16, 76
fifth dimension-Scale 203
final demise 141
finality 196
financial sector 105
finetuning 61
five cycles myth 73
flash crash 15, 187
flip-thinking 177
Fluxions 80
flywheel 39
fMRI 208
food industry 105
foresight reductionism 247
fossil energy industry 105
Four Growth Curves 132
Fourier analysis 73
four spheres 21
Four Styles of Thinking 194
fractal 218
Frankenstein scholar 199
free will 48, 249
French Revolution 219
freon 162
Freud, Sigmund 48
Fukushima nuclear accident 274
fungal culture 200

Gaia 116
Galilei, Galileo 42
game developers 253
game theory 244
gas laws 246
GDP 182
generalization 190
General Theory of Relativity 174
generatio spontanea 60
genetically modified genes 55
genimal 257
geology 47
geometric mean income 182
geosynclinal 47
GINI coefficient 121
glued-on sustainability 33
GMI 183
Go 253
GOFAI 254
Google Translate 255
Gossen's Second Law 77
grab as grab can 106
Grand Super Cycle 70
Grand Unified Theory 248
gravitational waves 174
gravity 42, 173
gray goo 143
Great Ape Project 115
Great Apes 115
Great Moderation 16
Great Recession 14
green GDP 160
Groningen 49
Gross World Product 89, 95
growth 93
growth is necessary 96

Gulf Stream 138
GWP 89, 95

Hailey and Huey 269
Harari, Yuval Noah 154
harmonic mean income 183
Harrisburg nuclear accident 272
Hawking, Stephen 198
HDI 262
Heisenberg's Uncertainty Principle 170
hindsight reductionism 247
HMI 183
Hobbes, Thomas 111
holism 196
holograms 217
homeopathy 191
homeostasis 233
homo economicus 76
homo neanderthalensis 112
homo sapiens 112
homo solitarius 115, 127
Hooke, Robert 208
Hothouse Earth 239
housing crisis 11
Hoyle, Fred 144
hubris 126
human capital 34, 125
Human Development Index 262
human dignity 124
human rights 78, 160
human species richness 115
human system inertia 164
Hume, David 111, 172
hybrid car 179
hypothesis 194
hypoxic areas 138
hysteresis 162
hysteria 47

ice ages 239
I, Me 170
immortal soul 231
impetus 44
impetus word 46
inclusive 226, 274
independent research 199
induction 175
inequality 116
inflation 187
inhibited growth 131
insight into who we are 109
institutional school 63
integrated chain management 30
intellectual output 91
interdisciplinary 204, 277
Intergovernmental Panel on Climate Change 138
International Monetary Fund 14
International Resource Panel 53
International Union for Conservation of Nature 115
intrinsic 224
intrinsic sustainability 33
Intrinsic Sustainability Goals 161, 205
invention 171
Invisible Hand Myth 74
IPCC 138
ISGs 161, 205
ISGs proposal 205
IUCN 115
ivory tower 199

J.P. Morgan 8
Juglar cycle 69

Kant, Immanuel 193
Kennedy, John F. 75
Kepler, Johannes 42, 65
Keynesian school 63
King, Mervyn 79
Kitchin cycle 70
knowledge theory 193
Kondratieff wave 69
Kuhn, Thomas 58, 217
Kuznets swing 69
Kuznets, Simon 75

lag effect 161
Lakatos, Imre 216
Lamarckism 60
land ice 138
last downfall 144
Lavoisier, Antoine 46
law 194
law(s): Clarke's three 37; collision 246; Comparative Benefit 77; Conservation of Belief 192, 211, 219; Conservation of Energy 170; Conservation of Life (after Death) 170; Conservation of Money 186, 264; Conservation of Money Current 264; Conservation of Things 169; economic 186; gas 246; Gossen's second 77; gravity 175; Kepler's three 65; mortgage 149; motion 42; patent 55; Roman 111; Roorda's first 37, 266; Say's 75; uncertainty principle 170
leaking Roof 158
Leeuwenhoek, Anthony van 208

left and right 179
Leibniz, Gottfried Wilhelm 80
Lenin, Vladimir 87
level fallacy 249
level transition 246
limit cycle 235
limits to Growth 98
Linnaeus, Carl 210
litmus test 208
lobbying industry 105
locally established economies 86
Locke, John 111
logic 193
logistic growth 130
logosphere 24
Lonesome Century 127
Lonesome Human 127
love 197, 249
Lovelock, James 116
Lucas, Robert 16
Lyell, Charles 61

magic 191
Magnetic Resonance Imaging 208
majority decisions 268
Malawi 117
Malthus, Robert 129
Man belongs to the Earth 56
manipulation 271
Mao Zedong 87
Marxism 78
Marxist school 63
Massively Multiplayer Online Roleplaying Game 252
maximum speed 178
Maya 172
MDGs 159
Meadows, Dennis 98, 133
medicine 47, 104
meme 26
Mendeleev, Dmitri 210
Mendel, Gregor 80
Mercury (planet) 174
methodical 222
Miasma theory 211
Millennium Development Goals 159
minimum sluggity 178
MMORPG 252
Mohawk Indians 101
money 12
money attracts money 119
money in 2110 188
Money Myth 78
monoculture 136
Montreal Protocol 162
Moon Treaty 53
Moravec's paradox 255
multilevel 224
multilevel networks 244
mutually reinforcing consequences 139
myth: Blessings of Capitalism 77; economics 90; economic growth 98; efficient market 76; five cycles 73; homo economicus 76; invisible hand 74; money 78; perfect market 75; revolution 78; rising tide 75; Say's 74; sustainable growth 99; Trickle-Down 75; Universal Economic System 85; value 78
mythosphere 25

Naked Credit Default Swap 9
National Academy of Sciences 193
Naturalism 61
natural system inertia 164
Neanderthal 112
neoclassical model 77
neoclassical school 63
Neptune 173
network theory 243
neuropsychology 48
neurotransmitter 197, 249
new-Institutional school 63
Newton, Isaac 42, 79, 173
Nigeria 51
noble savage 114
non-prime 15
North China Plain 139
nuclear energy, BSD on 272
nuclear lobby 274
nuclear power industry 105

object permanence 169
ocean currents 138
odyle 47, 231
OECD 16
omniconomic experimentation 261
omniconomics 3, 35, 41
omniconomy 34
optical illusion 67
orangutans 112
Organization for Economic Co-operation and Development 16
Orrell, David 79
Our Common Future 158
overexploitation 100
oversimplification 83
ownership 110
Oxfam 118

oxytocin 197
ozone layer 162

pandemic 136
paradigm 26
paradigm shift 34
pareidolia 67
Paris Climate Agreement 160
parliamentary democracy 268
participatory democracy 274
patent 55
patents on medicines 107
pattern recognition 64
peaceful 225
people 20
perceived temperature 140
perfect market myth 75
periodic table of elements 210
permaculture 30
personal message to the reader 279
pharmaceutical industry 103
phase diagram 235
phenosphere 26
philosophy 193
philosophy of science 193
phlogiston 46
photosynthesis 215
physicalization 80, 85
physics and astronomy 42
Piketty, Thomas 75, 121
Planet 19
Planet, Ltd 89
politosphere 26
Polo, Marco 244, 252
Popper, Karl 197
PopSim 131
populists 271
positive feedback 229
positivism 195
poverty 121
predators and prey 235
Price gouging 104
probability calculus 80
problem ownership 274
production per person 94
profit 20, 102
profit-seeking 102
Property 49
prosperity 20
protophysics 65
protoscience 59, 145
pseudoscience 60, 105
Ptolemy, Claudius 218
pure science 195
purposeful 226
PZM21 209

Qatar 117
Qatar Investment Authority 117
quality years 26
quantum biology 215
Quran 181

Raworth, Kate 30
real time strategy game 255
received View 195
red List of Threatened Species 115
reductionism 196, 246
referendum 270
reflective 222
Reis, Ricardo 79
religion 154, 192
replicability 223
replication 86
Republic of Groningen 49
research area of omniconomics 204
restasis 101
revolution 219
revolution myth 78
Ricardo, David 77
rightful ownership 112
rising tide myth 75
Roman Empire 141
roof tile structure 250
Roorda's first law 37, 266
Rousseau 114
Rousseau, Jean-Jacques 114
rubisco 215
Russian Revolution 219

Saint Regis Mohawk Indians 101
salami slicing 104
Sars-Cov-2 139
Say, Jean-Baptiste 75
Say's Myth 74
scale 203
scale dimension 245
scenarios 133
school of Chicago 77
schools in biology 60
schools in economics 61
Schumacher, Ernst Friedrich 30
Schumpeterian school 63
scientific criteria 222
scientific revolutions 58
scientific success 91
SDGs 159
sea level 137

Seattle, Chief 56, 159
securitization 11
self 172
self-learning system 253
self-reinforcing effect 229
sensitive soul 231
Serageldin, Ismael 19
simulations 234
singularity 143
Six Sigma 125
sixth extinction 136
slavery 126
Sluggity 178
small is beautiful 30
Smith, Adam 74
Snow White 171
social networks 126
social sustainability 20
societal output 91
soul 231
sources of the Nile 171
Spanish flu 139
SpazeGaze 106
Stalin, Joseph 87
statistical mechanics 246
statistics 80
stewardship 111
Stiglitz, Joseph 82
stopgap 218, 270
strategic 225
string theories 217
subprime 11
success 91
success score 262
Supply Chain Management 30
support 268
survival of the fittest 257
survival of the sustainablest 263
sustainability criteria 224
sustainability economics 153
sustainable development 20
Sustainable Development Goals 159
sustainable growth 99, 160
svalbard Global Seed Vault 28
symptom control 32
systemic criteria 223
system inertia 161

tegular unification 224, 250
television 171
temperature 246
Thaler, Richard 153
the dance 165, 278
theory 194
Theory of Really Everything 181, 194, 250
thought network 168, 243
tipping point 138, 238
tobacco industry 105
Toulmin, Stephen 209
transdisciplinary 226
transdisciplinary science 277
transfer 105
transparency 226
trickle-down myth 75
triple bottomline 19
Triple P 19
true prices 105
TV 171

Ukraine referendum 271
UNCED 158
understanding 176
UNESCO 159
unification 248
unifying 224
United Arab Emirates 117
United Nations 158
United Nations Conference on Environment and Development 158
Universal Atlas of Scales 203
universal economic system myth 85
upside-down thinking 177
uranus 173

value-free science 199
value myth 78
values-based science 200, 225
verification 197
Vienna Circle (Wiener Kreis) 195
virtual reality 256
vis essentialis 47, 231
vitalism 47
voting 268

Walras, Léon 76, 81
wave experiment 70
weakly inhibited growth 131
wealth tax 160
Wegener, Alfred 47
wet-bulb temperature 140
Wiener Kreis 195, 218
women's suffrage 78
World3 133
Wyndham, John 178

zoonoses 136

For Product Safety Concerns and Information please contact our EU representative GPSR@taylorandfrancis.com Taylor & Francis Verlag GmbH, Kaufingerstraße 24, 80331 München, Germany

Batch number: 10777124

Printed by Printforce, the Netherlands